PONDS AND LAKES

T. T. Macan

Naturalist and Deputy Director,
Freshwater Biological Association,
Ambleside

LONDON · GEORGE ALLEN & UNWIN LTD
RUSKIN HOUSE   MUSEUM STREET

Printed in Great Britain by
Cox & Wyman Ltd,
London, Fakenham and Reading
in 11 on 12 pt Times Roman type

PREFACE

Much of the book was written between Southampton and San Francisco on board the MV *Oronsay*, and it was completed at Idaho State University. Anyone writing away from his base must either carry with him an enormous weight of papers or rely on somebody to provide information on small points for which time could not be found before departure, on points in the literature which were not thought important originally, and on points which crop up unexpectedly on second thoughts and during revision. I chose the latter course and gratefully acknowledge the assistance of Miss Annette Kitching, who kept me supplied with a steady stream of information and never failed to find the answer to any question posed. It was also Miss Kitching who drew most of the diagrams.

Anyone working in a large laboratory devoted to one branch of science must be indebted to his colleagues for information, stimulating ideas and criticism, even though he has not deliberately sought any of them. Any Briton working in America becomes aware of how much at home he owes to a competent professional assistant staff. Lastly, anyone who can remember early days when he himself was in charge of the library is keenly aware of how much his work is facilitated by a good library staff.

CONTENTS

1 SETTING THE SCENE

Why study a pond?

'It's fun,' says *A*, the enthusiastic naturalist.

'It's in the syllabus,' snaps *B*, the disillusioned teacher.

'Well,' says *C*, the calculating careerist, 'I believe it might throw light on a problem which interests *X* and *Y*. They are two coming men, and if I can catch their eye I shall have friends at court in a few years' time when I am looking for a top post.'

'A pond presents a limited environment without a continual interchange of population with neighbouring biotopes and is, therefore, suitable for the study of principles,' enunciates *D*, the serious ecologist.

'World starvation is a real threat. In many tropical lands an important amount of protein is raised in ponds. We should study production in ponds wherever we can,' explains *E*, the scientist with a social conscience. His colleague in the hygiene department, where research over the last seventy years has helped to create the problem that worries *E*, might add that several carriers of disease inhabit ponds, and knowledge about small pieces of water anywhere may prove useful to fighting these diseases.

We can ignore *B* and *C*; *A*, *D* and *E* merit our attention. *E*'s practical approach is obviously important and it is the basis of the International Biological Program, which is nearing its conclusion at the time of writing (1971). When its findings are published, comments made here will probably be rendered out of date, but, at the moment, it is fair to state that calculation of production in a pond will only be possible after more work in the fields of *A* and *D*. Until this has been done as will be seen, the calculations must be based on certain assumptions which may introduce a substantial error.

Our characters *A* to *E* have been introduced to give an indication of the line taken in this book. They are today's figures and

9

their attitudes will determine the future, or at least the imme-
diate future. To understand how the situation today came about,
it is necessary to look back into the past. It is fashionable to go
back to Aristotle, and to leap from his time to the Renaissance,
but for practical purposes biology started with Linnaeus some
two hundred years ago. He classified plants and animals and
gave each species a name. Until this had been done, little
advance was possible because nobody could be certain what
plant or animal an earlier investigator had been studying. The
nineteenth century saw great elaboration of Linnaeus' work and
considerable advances in the field of 'natural history', that is,
the study of what an animal did. This was almost entirely the
work of amateurs. One notes, for example, that the first professor of Zoology at Cambridge was appointed in 1866 though
there had been a professor of botany since 1825. The nineteenth
century was of course the heyday of a leisured class, and, for
those with a taste for science but without the fortune to indulge
it, there were few alternatives to taking Holy Orders and seeking
the serenity of a country parsonage. The Reverend J. G. Wood
was a prolific writer of natural history books, the Reverend A. E.
Eaton wrote a monograph on the Ephemeroptera which is still
consulted today, and some of the early studies of fossils eman-
ated from rural vicarages. They probably represented our
character A although perhaps, if they had been asked what in-
spired them, they would have replied that they were seeking to
disclose the wonders of the Creator.

In 1859, that notable amateur, Charles Darwin, published the
Origin of Species. This set the scientific world an intellectual
problem; if complex plants and animals had evolved from
simpler ones, it should be possible to work out the family tree
of the plant and animal kingdoms, and to base the Linnaean
system of classification on relationships. To this task zoologists
and botanists bent themselves with zest, and their sciences
became a study of morphology, of comparative anatomy and of
embryology. When the writer was a student, in the early thirties,
he studied in a laboratory of 'zoology and comparative anatomy' and to this day wonders why a distinction was made.
However, the staunch pillars of this school were then at the end
of their time and, when they went, a new generation took over.

10

This was the generation that had taken part in the First World War and was imbued with the new ideas that always emerge from the upheaval of war. They found that dry bones made zoology a dry subject and they deplored the way in which the older morphologists named structures according to what they thought they did without taking any steps to find out. For example any thin-walled extrusion from the body surface was called a gill; subsequent work has shown that many of these structures have no respiratory function.

Zoology changed suddenly. Comparative anatomy and taxonomy fell out of fashion and experimental zoology, which is physiology, came into vogue. A knowledge of physics and chemistry became more important than a knowledge of nerves and blood vessels. Ecological and taxonomic studies were discouraged as subjects for Ph.D. theses on the grounds that the work was purely descriptive and the intellectual content was small in comparison with work that involved experimenting. 'Natural History' came to have a pejorative meaning.

Not much work on fresh water had been done in Britain before the First World War apart from Victorian studies of the habits of individual species. The enemy submarine campaign during it caused attention to turn to inland fisheries as a possible source of food, and this led to a general interest in freshwater biology as did also the report of a Royal Commission on sewage. Interest in animal ecology in general was stimulated by the publication of Elton's *Animal Ecology* in 1927. The stage should have been set for rapid advance but advance, in the author's opinion, was possible only in three phases. The first was taxonomy. This was particularly important. Most groups had been thoroughly examined by taxonomists when taxonomy was popular, but descriptions had been largely confined to the adults. Some insects have a long aquatic stage and a very short aerial adult stage, and therefore it is the aquatic stage that the ecologist encounters most often. It was to this stage that he needed keys that did not exist. The second phase was a description of the communities in the various biotopes (a biotope is used here to describe a particular kind of ecological condition such as open water, sand, or stony bottom). Finally comes the third phase when experiments are planned to explain what has

11

been described. Unfortunately at this time the first two phases had fallen out of fashion.

Such work as was done bears the unmistakable print of the current ideas and shows evidence of a belief that the distribution of species could be explained in terms of physics and chemistry. Any study of a pond involved extensive chemical analysis and careful recording of temperature. The idea that the animals and plants themselves might be the most important factors in the environment has been gaining ground only recently.

The year 1970 was European Conservation Year, and a great deal was written and talked about the environment. A few years before, the system of allocation of grants for research had been reorganized, and the Natural Environmental Research Council had been set up. The first event may have had some influence on thinking in the academic world; the second certainly has. Prophecy is always rash, but it does seem likely that a more logical investigation of the natural environment, particularly fresh water, because it is so conveniently polluted and so important for many purposes, will be attempted.

So far this study of the development of research has been confined to zoology. The history of botany has been more straightforward. Apart from the fungi, plants can, by definition, only feed in one way, in contrast to animals whose source of food is varied. The biggest plants tend to dominate, whereas the biggest animals must of necessity maintain populations that are small. Plants in consequence are less diverse, and plant taxonomy has always been years ahead of animal taxonomy; for example the well-known flora of Bentham and Hooker first appeared in 1858. The inability of plants to move has been another factor facilitating the study of their ecology, and it is no surprise, therefore, to find that this was moving forward steadily in the early years of this century. The composition of the various communities was described, and the conditions under which each one was found were noted. These studies culminated, in so far as studies of this kind can culminate, in 1939 when *The British Islands and their Vegetation* by the late Sir Arthur Tansley appeared. Now, an important aspect of botanical ecology is experimentation to find out why communities are constituted as they are. If the practitioners ever wonder whether their zoological colleagues have not sometimes put the cart before the horse, there are some grounds for their speculations.

Some applied research has been mentioned but a little more must be added as a footnote to this introduction. Scientific interest in the cultivation of fish in ponds was shown on the continent just before the end of the last century. In Britain coarse fish are not regarded by most people as fit for food except in war-time, but today they are attracting official recognition because angling is such a popular recreation. The turn of the century was also the great period for unravelling the life histories of parasites that cause disease in man. Some of them are carried by animals that have a stage in fresh water, notably the mosquitoes. Considerable investigation of ponds followed the incrimination of these insects as vectors of disease.

This book is written for those who have not reached the research stage and perhaps do not intend to. It is planned upon the following ideas: the study of a pond must start with some knowledge of what occurs in it, even if it does not extend beyond the commonest species or the species in one group. A knowledge of numbers is the next advance and this leads on to a study of life histories and of where the individuals are at different stages of development. Full information on these points is what the student of production seeks. All along the line the question why arises; why, for example, is any given species distributed as it is? It is necessary, when writing, to keep all the balls in the air at once, a feat which, practically, would tax a team of expert investigators. An attempt has been made, accordingly, to indicate where and how a single ball can be isolated.

2 THE CYCLE OF MATERIALS IN LAKES

SUMMARY

Cyclical processes, powered by solar energy, are possible in standing water. There is also generally an input and output of matter. Substances washed in from air or soil are used by plants in the creation of living matter. Whichever is in shortest supply determines how much is produced. Living plants produced in the lake, together with much vegetable material of terrestrial origin, are eaten by animals. Dead material, from within the lake or without, decomposes and releases substances for cycling. In the summer many lakes are divided into cold lower and warm upper regions, which impedes cycling. Within the two regions the water circulates.

A lake is a bulge in a drainage system, where the flow is so slow that certain processes that would not otherwise be possible can take place. This is true in Britain but in some parts of the world, Iran for example or Utah in the USA, rivers end in lakes, and the water they bring in leaves again as vapour. This process soon produces, needless to say, a highly saline lake. The opening statement implies that a lake is a fairly large body of water. Smaller ones may be no more than holes in the ground, not obviously part of any drainage system except after rain. In a rock bed they may be isolated, but in many parts of the country there are hollows where the soil surface, owing to a natural dip or to excavation, lies below the surface of the water table. Open water may appear stagnant but is in fact part of a sheet that is moving steadily towards the nearest stream or river.

The processes referred to are cyclical, and the term cycle and cycling will be applied to them, leaving circulation for movement of the water itself. A cycle needs energy to keep it going round, and this is supplied by the sun. It is possible to imagine this cycling going on in a completely isolated lake basin. Living

14

material is built up from inorganic constituents with the energy derived from the sun. It dies and decomposes into the original components which are then available for the start of another cycle. This is obviously a theoretical concept, for, if there is rain, material must enter the system from outside and, if there is no rain, it must dry up. Whether it really takes place is being tested by some of the author's colleagues at the time of writing. They have isolated part of a lake by means of an enormous polythene cylinder whose lower open end is sunk in the mud, and whose open upper end is supported by floats above the surface of the water.

The result of the building-up half of the cycle is commonly referred to as production, and this term is used here. Some restrict it to the activities of plants, which are the only organisms that can create living from inorganic material. The animal kingdom, they maintain, cannot produce in this sense but only convert, directly or indirectly, the material built up by plants into its own flesh. The author's colleague, Mr F. J. H. Mackereth, has pointed out that, whatever may be possible in a lake, in global terms and in the long run processes must balance and the final product is nil. If the building up were to exceed the breaking down, the world would become overpopulated, while the reverse would lead to the disappearance of life. In this connexion it may be pointed out that capital accumulated in the past, notably during the Carboniferous period, when dead bodies of animals and plants were preserved in fossil form as coal and oil, is making possible the current trend towards human overpopulation.

Standing water may also be described as nature's dustbin. In temperate regions a vast number of dead leaves are shed at the end of each growing season. These come to rest in many places, but wind in a dry spell may set them on the move again. Permanent water, however, is a final resting-place. A small pond will trap many, a lake relatively more because great quantities are brought in by the inflowing streams and rivers, but this is not all that the inflows bring in. To begin with rain is not pure water. A storm at sea will sweep into the air droplets which may come down in rain far inland, and clouds that have travelled across land will shed rain in which there is a variety of substances derived from smoke and other forms of atmospheric pollution. The rain may seep through the soil before reaching a lake, dissolving much from it, and also modern methods of water-borne sewage disposal add considerably to the substances present. The first stage in the study of a lake or pond is, therefore, primarily the domain of the chemist, but much may be learnt from an intelligent study of the drainage area. If the soil is predominantly rocky or so poor that the farmer can use it for nothing except the grazing of sheep, each of which may require more than an acre, the amount of life produced in the water will not be great. If the soil is rich and the farmer's production on it is obviously high, production in the water may be expected to be high too. Human settlement will raise production.

The study of a piece of water thus begins in the drainage area. The next stage is the study of the plants. In a large deep lake most of the primary production is by the phytoplankton, the minute algae suspended in the open water, and the contribution by the plants rooted in shallow regions, and by the algae growing on them and on rocks or stones is slight. The reverse holds in a shallow lake. Plants cannot extend far down because light is quickly absorbed by water itself and by anything organic or inorganic in suspension. There is no growth of plants below about six metres in Windermere and the limit may be well above this in rich ponds.

Plants synthesize water and carbon dioxide into sugar:

$$6 \ H_2O + 6 \ CO_2 = C_6H_{12}O_6 + 6 \ O_2.$$

As mentioned above the sun supplies the energy. The energy is released when the sugar is burnt, as it is when animals, or plants themselves, respire. Further elaboration to form proteins requires the presence in available form of various elements, which are derived from sources discussed above. The amount of vegetation produced may be limited by the amount of carbon dioxide, or by the amount of energy available, but it is generally limited by the amount of one of these other substances, which for convenience may be called nutrients. Almost all the ions dissolved in fresh water are contributed by sodium, calcium, magnesium, potassium, hydrogen, bicarbonate, chloride, sulphate and nitrate, which are present in very small amounts. In

16

Windermere, a soft-water lake, not one of them attains a concentration of ten parts per million, though in hard waters there is a good deal more than this of calcium and bicarbonate. These are the elements whose concentration is generally measured, but before this is undertaken, it is as well to pause and consider what the purpose of the analysis is. Not everybody has done this. The waterworks engineer attempting to produce a potable water is concerned with one set of criteria, the ecologist searching for a factor limiting distribution with another. The important element for the investigator of production is the one in shortest supply. Of the nine mentioned above only nitrate is commonly the limiting factor. It and phosphate are the main constituents of fertilizers added to the land by farmers to increase production. Potassium is often added as well, but in the Lake District lakes, it and the other ions listed above except nitrate remain so constant in concentration throughout the year that they are obviously present in amounts far in excess of the needs of the plants.

Many of the algae are diatoms, which produce a siliceous skeleton like a box and a lid. One of these, *Asterionella*, reproduces rapidly in Windermere when conditions, principally light, become favourable in spring and Dr J. W. G. Lund has shown that it is exhaustion of silica which brings this reproduction to an end. Dr C. R. Goldman in America has cultured different algae in lake water and added various substances. Several have been found to increase production and these are assumed to be limiting in nature. Some are present in very small amounts. There is evidence too that certain organic compounds may be necessary for growth and reproduction. All this work involves substances present in minute quantity and is not feasible except in well-equipped laboratories.

These plants are eaten by some animals, but dead leaves and other material produced outside the lake may be important at the base of the food chain. Animals have to establish themselves in a place where they can obtain food and at the same time find safety from various adverse factors, chief among which are other animals which prey upon them.

The whole process of production is influenced by various physical events in a lake, both directly and indirectly, through chemical changes. The sun's radiation, as has been stated, does not penetrate far into water, and that part of the spectrum which warms water is absorbed most rapidly. On a hot still day only the layers immediately below the surface rise in temperature but wind mixes this water with the colder layers below. However, there comes a time when the difference in density between the warm upper layers and the cold lower regions of a lake is so great that the fiercest storm will not mix them. They remain separate for the rest of the summer, the upper part becoming warmer, the lower part remaining at a temperature only a degree or two above that of the winter. As the sun declines in the autumn and early winter, the upper layers lose heat and eventually, perhaps not until December, a storm mixes the whole lake. Further cooling leads to a uniform temperature from top to bottom of 4 °C, at which temperature water attains its greatest density.

Production is largely confined to the upper warm layers because only there is enough light. Many of the animals and plants that die sink to the bottom of the lake, decomposing there and on the way, into their basic components. But these cannot return to circulation because the cold layers are cut off from the warm upper ones. In winter when storms mix the lake and rain falls on the drainage area they are washed out and carried down to the sea.

This decomposition, effected by bacteria and fungi, uses oxygen. If the volume of the lower cold part of the lake is small, and if there are a lot of dead leaves there and many dead bodies descending from above because production in the upper layers is massive, all the oxygen may be used. An oxygen deficit must obviously remain until the waters mix again. Early in the present century, the adjective 'eutrophic' was applied to such a lake. Today 'eutrophication' has passed into everyday speech and is applied to any enrichment of a lake due to sewage effluent or agricultural fertilizer. If a lake is less productive, and the basin deeper, two features which often go together, supplies of oxygen in the lower layers are not exhausted and the adjective 'oligotrophic' is applied. Fig. 1 shows what is commonly called the 'profile' of two lakes. In both the temperature drops suddenly in the region known as the thermocline and the shape of the two

18

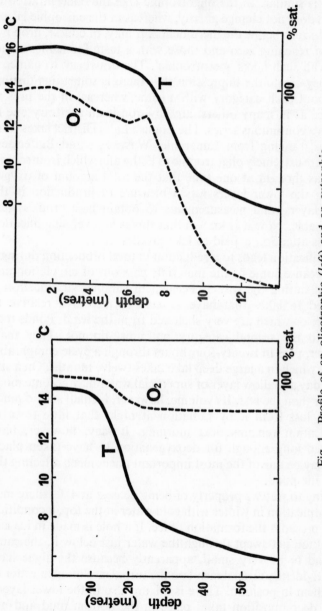

Fig. 1 Profile of an oligotrophic (*left*) and an eutrophic (*right*) lake in summer. (After V. G. Collins, 1970, *Water Treatm. Examn.* 19.)

curves is similar. In the oligotrophic lake the concentration of oxygen does not change greatly, whereas in the eutrophic lake it drops to zero below the thermocline. It may, of course, drop but without reaching zero and those with a taste for names sometimes call such lakes 'mesotrophic'. The drawback to names is that they create the impression that there is something distinctive about each category with a name, whereas, in the present instance as in many others, any division is an arbitrary one in what is a continuous series. The English Lake District lakes form a series, ranging from Ennerdale, Wastwater and Buttermere which are extremely oligotrophic to Esthwaite which is eutrophic.

It was thought at one time that the total amount of oxygen used in the lower layers was a measure of production in the upper layers, and measurements to obtain lake profiles were fashionable. Now it is known that this is an over-simplification and less attention is paid to lake profiles.

Stratification leads to a reduction in total production during a year because some of the materials pass out of circulation and are lost during the winter. Permanent summer stratification is confined to lakes that have a small surface area relative to volume, or which are very sheltered from the wind. Ponds may fall into the second category but most do not. They may, however, pass in twenty-four hours through a cycle comparable to that which in a large deep lake takes twelve months. On a still sunny day a shallow layer of superficial water may become much warmer than the rest. Its volume, however, is small and so much heat is lost from it by radiation at night that little trace of stratification remains next morning. It may, however, have persisted long enough for deoxygenation to have taken place. This may be one of the most important phenomena affecting the life in the pond.

Owing to water's property of being densest at 4 °C, there may be stratification in winter with cold water at the top, a condition which precedes the formation of ice. If a hole is made in ice and a plankton net swept through the water just below it, the catch is found to be very small, apparently because the algae have sunk. Evidently algae depend on the turbulence of open water to keep them in position. There is mixing also in the lower layers. Much decomposition takes place in the bottom mud and nutrients are released and diffuse away. But diffusion is an extremely slow process, and it had been known for many years that the lower layers are much more uniform than they would be if diffusion alone operated. Dr C. H. Mortimer became interested in this problem during his classic study of the exchange between mud and water in the English lakes. Wind can account for the mixing of the upper layers: even a light wind will set the surface layers of a lake flowing and there must be a return current, generally along the top of any temperature discontinuity that may exist. What mixes the lower layers is less easy to see, and Mortimer decided to follow up some work that had been done during the bathy-metric survey of the Scottish lochs carried out in the first decade of the present century.

His first venture was to sit for twenty-four hours in a boat moored in the middle of Windermere and measure periodically the temperature of the water at various depths. This was tedious, and, without a team, did not yield the continuous observations that were clearly essential. Accordingly he slung a series of thermistors from a buoy and connected them to a recorder in the laboratory. It was then possible to read the temperature at nine different depths in the middle of the lake at any desired moment. The picture was as described: warm water down to a depth of between five and ten metres, then the thermocline, the layer in which the temperature dropped rapidly, and finally the bottom layer at 6 °C. However, if a record was examined after the passage of time, it was clear that the thermocline was moving up and down.

Now this investigation is well beyond the scope of any but a limnological laboratory. Several kilometres separated the ther-mistors and the recorder, and between each one there had to be a wire. However, as a result of this work Mortimer devised a model, by means of which he could demonstrate the course of events in a lake and this model could easily be copied by anyone with access to a moderately equipped laboratory or workshop.

Mortimer built a section of lake roughly of the shape shown in Fig. 2, and enclosed it between two sheets of plate glass 3 inches apart. His 'bedrock' was actually of concrete but this made the model unnecessarily heavy and today it would be easier to use fibre-glass. The basin was 5 feet long and 10½ inches

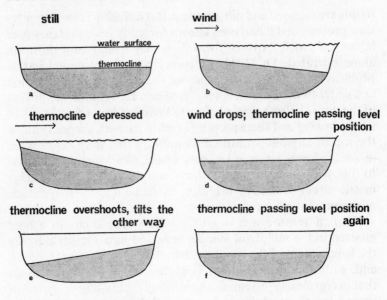

still

wind →

water surface

thermocline

a

b

thermocline depressed →

wind drops; thermocline passing level position

c

d

thermocline overshoots, tilts the other way

thermocline passing level position again

e

f

Fig.2 Seiches in a lake.

deep, and the whole model was 6 feet long, the additional foot being taken up by the frame, and a control unit.

The tank was filled with cold water to represent a lake at the end of winter. The sun's heat was provided by two wires through which an electric current could be passed. When a suitable temperature had been reached, methylene blue was carefully run in to the warm water to distinguish it from the cold water below. The model was now in the state of a lake on a still day in mid-summer (Fig. 2a). Anything from a gentle breeze to a gale was provided by two ladies' hair driers controlled through a rheostat. Water is heavy and a strong wind tilts the surface of a lake the size of Windermere so little that only elaborate apparatus can detect it. The difference in density between the two layers at different temperature is not great, and it is obvious that, after the wind has been blowing for a while, the blue water is deeper at the leeward end and shallower at the windward. In other words the thermocline is tilted (Fig. 2b, c). At full blast the hair

22

driers will drive the upper warm water so far to leeward that the cold water is exposed at the windward end, and this phenomenon has been observed in real lakes. Under these conditions the blue water is very disturbed but it remains separate from the rest in a most convincing way.

When the wind dies down, the warm water piled up at the leeward end rolls back, but its momentum carries it past the point at which the thermocline is level (Fig. 2*d*) and the thermocline tilts in the opposite way (Fig. 2*e*). Then the upper layers return and overshoot once more, though each time the angle of the thermocline is a little less till finally the see-saw stops. In Windermere a wind generally springs up to impose a new pattern before the old one has disappeared. An oscillation takes 18–19 hours in the lake, but in the model the whole process can be demonstrated in as many minutes. It is obvious that the thermometers near the top of a line slung midway between the pivotal point and the end of the lake would be bathed alternately in warm blue water and cold colourless water. The results obtained by Mortimer with his thermistors in Windermere are easy to understand almost at a glance.

The conclusion of the demonstration of the model was the release of a few crystals of potassium permanganate. As each one sank to the bottom, it left a column of coloured water, and from this it was evident that the oscillation of the thermocline was accompanied by streaming in both the upper and lower layers. This flow impinging on any irregularity in the basin sets up eddies, and in this way the water is mixed.

There have been many attempts to distinguish scientifically a lake from a pond. One criterion is the establishment of stratification throughout the summer, but this depends on the shape of the basin. For example Lough Neagh has the largest surface area of any body of fresh water in the British Isles but, because it is relatively shallow, it does not stratify. According to this definition it is a pond, which is ridiculous. Another definition of a lake, as distinct from a pond, is a piece of water so deep in places that there is not enough light for plants. This is not satisfactory because a small shallow pond fouled by cattle may support such a dense population of algae that the light is absorbed in a very short distance. A third school defines a lake as a piece

23

of water of such size that wave action erodes the shores. However, some moderately large bodies of water are fringed with reedswamp too well established to be torn up by the most violent storms. A definition that is watertight may be found in a limited area but the author's view is that, in general, it is misleading to draw arbitrary lines across a continuous series.

Ponds are much less stable than lakes. Heavy rain may change completely the water in a pond. In dry weather it may disappear. One still warm night when a big crop of algae is dying may lead to deoxygenation. A chance visit by a few ducks may produce considerable alteration to the vegetation. The larger the piece of water, the greater its stability, and here too we are faced with a continuum that can only be divided arbitrarily.

Few who wish to study water will be unable to find some small body of it, a pond in fact, within convenient distance. It is to be regretted that ponds have been neglected by professionals, who have generally worked on the largest lake they could find, but some may look on this as a challenge rather than as a matter of regret.

3 CLASSIFICATION AND ADAPTATION

SUMMARY
Animals can be classified according to where they live in relation to the open water, according to their method of feeding, or according to their life history. Some animals have passed from sea to fresh water and overcome the physiological problems created. Most entered fresh water from the land, and have subsequently modified their ways of moving, respiring or feeding.

Chapter 2 introduced the basic processes; the present chapter introduces the animals and plants. There is no space for a long account of the various groups separated according to their structural characteristics, and the reader is referred to the well-tried works of Mellanby (1963) and Clegg (1965). Here some functional classifications will be discussed.

Animals fall into reasonably distinct categories according to whether they live:

on the surface,

in the open water,

among weeds or stones, or

in the mud.

This is another familiar system described in many books. Plants can be classified in the same way. The surface of a small sheltered pond may be covered by *Lemna* the duckweed, or by *Azolla*, a small fern. (In tropical waters floating higher plants may be a serious pest.) Algae, generally unicellular, abound in the open water. Others attach themselves to stones, and it is on this substratum that the familiar filamentous forms are found. Mosses also grow attached to a hard surface. Sand and finer soils are colonized by the rooted higher plants, and also by mosses and various other simple groups such as the stoneworts and the quillwort *Isoetes*. These plants can be subdivided into those that are permanently submerged except sometimes for a small inflorescence, those with floating leaves such as the water-lilies, and those which emerge from the water, such as reed, reedmace, bulrush and sedge. These succeed each other as a piece of water gradually fills in.

Life in the water is a constant search to secure food without becoming food, and the pattern imposed by this rigorous struggle calls for examination. Plants cannot flee and must occupy a position in the light, but it is rare to see higher plants extensively damaged by grazing. They seem to have acquired some sort of immunity but whether it is by means of hard cuticle, impalatability or other method is not known. Algae are eaten extensively, though some, protected by a cuticle, may pass unharmed through an animal's alimentary canal. Snails have a gizzard full of fine sand in which algae are triturated so that the cell-wall is ruptured and the digestible contents within exposed to the gastric juices. Algae are small, which means that anything feeding on them takes time to secure a full meal, and that they can reproduce rapidly. The effects of predation are reduced in this way by some animals too.

Larvae of some insects live in the tissues of higher plants and feed on them. They are possibly the only pure herbivores. Snails, the nymphs of certain Ephemeroptera, and various other animals browse on the felt of algae that covers the surfaces of higher plants and stones, but, as anything falling from above may become entangled in this felt and as small animals live in it, their diet must be a varied one. Many animals feed on detritus, which consists of small pieces of disintegrating plants and animals, as well as inorganic particles. It is not too difficult to discover what they take in but much harder to find out what they can make use of. Enzymes that can break down cellulose are found mainly among bacteria and fungi and it may be that it is these living organisms that are the main source of food for animals that feed predominantly on plant debris. The exploration of this field is in its early stages.

Some of these detritus feeders walk or swim in search of food. Others create a current to bring fine particles to them and can live in one place, which has advantages in the struggle for sur-

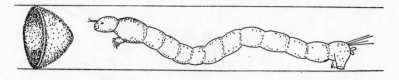

Fig. 3   Chironomid larva in tube. (Redrawn from B. M.
Walshe, (1950), *J. Exp. Biol.* **27**.)

vival against predators. Most of these lie buried in mud. Dr
B. M. Walshe has described a chironomid larva which con-
structs a U-shaped tube and then spins a net across it (Fig. 3).
By undulating the body the larva drives water through the tube
and the net retains fine particles of debris. When the net is
clogged, the larva eats it and spins another. The lamellibranchs,
which range in size from the small pea-shell (*Pisidium*) to the
large swan-mussel (*Anodonta*), keep a current flowing through
their bodies by means of innumerable cilia, and trap particles in
a thread of mucus which passes into the alimentary canal.

The carnivores may be divided into those that lie in wait for
their prey and those that chase it – the lurkers and the hunters.
All the dragon-fly nymphs are lurkers. The labium, the lowest of
the mouthparts, is a hinged structure which can be projected in
front of the head (Fig. 4). The passing of a small animal within
range stimulates this reaction and the labium is provided with
two strong teeth with which the prey can be gripped. This process is easy to watch in an aquarium. Another sessile carnivore
is *Hydra*, which hangs with long tentacles trailing in the water.
These are armed with cells which on contact eject a harpoon-
like structure bearing a substance that paralyses a small
animal into which it is injected. Any such animal which blunders
into the tentacles of *Hydra* is quickly rendered inert and then
conveyed inside the polyp's body.

Caddis larvae of the family Polycentropodidae spin funnel-
shaped nets and lie in wait in the apex ready to pounce on any
animal that collides with them.

Notable hunters are the beetles, both as adults and larvae. The
disadvantage of this method of obtaining food is that it renders

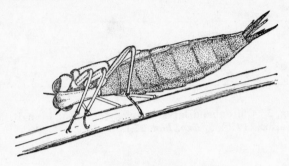

Fig. 4 Dragon-fly nymph (*Aeshna*) partly from below, showing mask. (Drawn by C. Joan Worthington.)

the practitioner in full chase liable to become a victim of a larger carnivore. As will be described later, water beetles were conspicuous in a tarn after all the trout in it had been removed and were seen no more when the tarn was restocked. They were found inside the trout. All our indigenous fish are carnivores and some are typical hunters. The trout in still water is a good example. In running water, however, it tends to lurk under the lee of a stone and dart up to seize organisms which the current carries past it, so to a certain extent it is a lurker. These terms, like so many others, are useful up to a point but misleading if applied too rigidly. Little is known about the exact feeding pattern of any freshwater carnivore, though a certain amount has been deduced about that of fish from what is found inside them. In Windermere for example trout and perch when small, and eels feed mainly on invertebrates. The perch evidently spends some time grubbing in the bottom mud and feeding on animals which it uncovers there, though it does not do this to the same extent as the carp. It does not feed at the surface which the trout does. It seems that neither of these two swimmers is willing to penetrate the dense forests of pondweeds and seek food there, whereas the eel, a much longer fish capable of moving after the manner of a snake, does exploit this rich source of food.

Some of the caddis larvae are carnivorous but, encumbered by a case, they cannot pursue prey at speed and have no modification to seize anything that swims unsuspectingly past them.

Dr H. D. Slack has recorded that in Loch Lomond larvae of *Phryganea* eat great numbers of the eggs of the powan (*Coregonus*), and presumably they feed at all times on animals incapable of swift flight. He who undertakes to examine what is inside a carnivore must have a good knowledge of the animals occurring in the same water, for he must be able to establish specific identity from heads and legs and other fragments. This sounds a formidable undertaking, but anyone familiar with the fauna of a small body of water will soon find it easy to identify the victims of carnivores from well-chewed or digested remains. Observation in the field is difficult because one of the facts that strikes an observer peering into a pond, at least one inhabited by fish, is the apparent absence of animals. A net passed through the vegetation, usually a veritable forest in relation to the size of the animals, then reveals an astonishing number and variety. Observation in an aquarium is a third possibility but it is not easy to ensure that conditions in an aquarium are exactly the same as those in nature.

Many insects that fly alight on the surface of the water and are trapped by the surface tension. At times trout feed extensively on these unfortunates. They are the main source of food for the water-striders (*Gerris*, Hemiptera) and the whirligig beetles (*Gyrinus*, Coleoptera). Another surface-dwelling water-bug, *Hydrometra*, has a very long head inside which are sheathed five stylets. Dr G. A. Walton has described how these stylets are used to spear waterfleas and other small creatures swimming just below the surface. An aquatic bug, the backswimmer or water-boatman (*Notonecta*), also feeds on surface casualties, though it lives in, not on, the water. It hangs from the surface by the tip of its abdomen, the paddle-like hind legs spread out ready to propel it to any prey whose struggles have set up vibrations in the water, its fore and mid legs ready to seize the prey and impale it on the sharp beak, through which, like all bugs it sucks the body of its victim (Fig. 5).

Life history is a third criterion by which animals may be classified. As will be shown later, it is of enormous importance in the economy of a community, and it has two advantages in the present context. The first is simplicity; when generations overlap, the unravelling of a life history may be difficult, but if

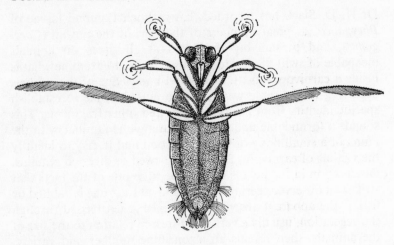

Fig. 5 Water-boatman (*Notonecta*) from above. (Drawn by C. Joan Worthington.)

they do not, and many do not, all that is required is a series of measurements of a sample taken at appropriate intervals. Measurement to the nearest millimetre is generally sufficient, and this can be done quickly if the animals are placed in a transparent dish over graph paper. Head width is more accurate but total length, excluding appendages, is generally sufficient. It may not be necessary to measure every specimen, the length of the largest, and of the shortest and what appears to be the most common length proving adequate. In parenthesis it may be remarked that the importance of accuracy in science is rightly emphasized to students, but they are not always instructed to think for themselves about what any set of measurements is designed to show and what degree of accuracy is desirable. The second advantage is that a life history may vary according to conditions, and therefore somebody who repeats what has been done elsewhere may not be treading a path that has been signposted already. Indeed the possibility of making a comparison may be an advantage. For example Fig. 6 shows the life history of *Rhithrogena semicolorata*, an ephemeropteron confined to streams. It has one generation a year, grows throughout the

30

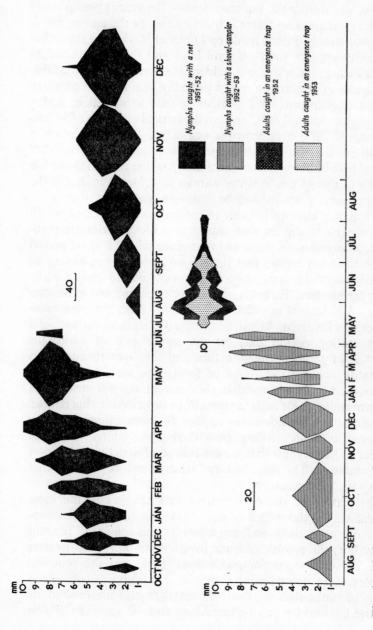

Fig. 6 *Rhithrogena semicolorata*. Numbers of nymphs caught each month in a net (*above*) and a shovel sampler (*below*). The histograms show absolute numbers caught arranged in millimetre size groups. The horizontal figure shows numbers of adults caught in an emergence trap in two successive years. (From T. T. Macan, 1958, *Verh. int. Ver. Limnol.* **13**, p. 846.)

Nymphs caught with a net
1951–52

Nymphs caught with a shovel-sampler
1952–53

Adults caught in an emergence trap
1952

Adults caught in an emergence trap
1953

winter, and during the warmest time of the year is present only in the egg stage. This figure is from a study by the author and it followed closely on one made by Dr Janet Harker. In this other stream emergence went on until later and the eggs started to hatch earlier; the water was colder. This led me on to examine emergence in one stream over a period of years and to compare it with temperature. Hot weather seemed to curtail it, and a temperature around 16 °C appeared a critical one which *Rhithrogena* could survive only in the egg stage. This is of course a hypothesis that must be tested but the ultimate test will be easier the more field observations there are. Anyone recording the life history of this species in water warmer or colder than that in the streams already studied will be making a useful contribution.

Before any attempt is made to categorize life histories, it will be profitable to digress and discuss how life histories are regulated. Many animals mate and reproduce within a short period and it has been argued that there is an advantage in having all individuals at the same stage at the same time to ensure maximum fertilization. However, some species mature and mate over a period of several months, and evidently there are other considerations involved. As will be seen later, the disadvantages of a short mating period may be the appearance of enormous numbers of young at the same time and consequently enormous mortality owing to shortage of food. The survivors may be surrounded by more food than they can eat. Several methods of circumventing this disadvantage will be described in due course. An important consideration applies to those insects that are airborne as adults. Many grow during the winter but there appear to be very few that are capable of adult activity at winter temperatures. The life history must therefore ensure that emergence is in summer.

The feature of the environment that changes in the most regular way is the length of day. This determines the development of the gonads and consequently the spawning time of many fish; the growth of some invertebrates is related to it as well. There is scope for much work here, and the practical problems are comparatively simple. The requirements are a series of tanks shut off from the daylight and illuminated by electric light which can be turned on and off automatically to give hours of illumination different from those outside. There must be some provision for the control of temperature.

Temperature varies in a much less regular way in a climate such as that of the British Isles, but some life histories are none the less related to it. Those of various dragon-flies, studied by among others Dr P. S. Corbet, joint author of a book on the group (Corbet, Longfield and Moore, 1960), serve to illustrate this topic particularly well. *Lestes sponsa*, the green lestes, lays its eggs in July and August, and development proceeds apace at the temperatures prevailing. However, before the embryo is ready to hatch, a period of what is known as diapause development begins. The feature of this is that the optimum temperature is low and not high. When diapause development is completed, high temperature is optimal once more. The effect of diapause is to ensure that no eggs hatch before the time when the temperature has fallen too low for hatching and that the winter is spent in the egg. Hatching takes place in April when the temperature has reached a suitable level, and thereafter the growth of the nymphs is rapid in comparison with that of other dragon-flies, being completed in three months or less. The rate of nymphal growth is determined by temperature as far as is known. This is a rigid type of life history.

Adults of *Anax imperator*, the emperor dragon-fly, emerge between mid-May and mid-July and the eggs hatch in about three weeks. The larvae grow rapidly and reach about half their full size by October when growth stops, probably because the temperature is too low. Growth is resumed in the following April and a few nymphs reach the final instar in May. If they reach this stage when daylength is still increasing at a rate of three minutes a day or more, there is no halt to development and adults emerge quite soon. If they reach the stage later, nearer the solstice, at a time when daylength is increasing at a slower rate, there is a diapause in the final instar and development is retarded until the temperature drops. Any nymphs that have been growing more slowly now catch up the large ones and in the following year all emerge within a short period of time. The total length of the period is six weeks but 90 per cent of the dragon-flies emerge during the first ten days. This is a more flexible life history than that of *Lestes*.

More flexible still is *Pyrrhosoma nymphula*, the large red damselfly. Mr A. E. Gardner saw a couple laying eggs on 6 June, and took some of them home to rear. The eggs hatched on 24 June. A month later the single survivor was 5.5 mm long and by 23 September it had reached a length of 10 mm. Growth continued up to Christmas by which time the nymph had attained the final instar and a length of 18.5 mm. There was then a pause in development. Eventually a length of 20 mm was attained and the adult emerged on 19 May. On its first birthday it could well have been reproducing.

Mr Gardner's eggs were collected in Hampshire. The author has made extensive observations in the Lake District. It is possible that a small percentage of specimens complete development there within the year, as just described, but it is difficult to be certain of this when a wild population is sampled, because growth as quick as this would soon take a specimen to a size at which it was indistinguishable from the laggards of the previous generation. Most eggs probably take longer to hatch than those observed by Mr Gardner, for small larvae were not abundant before September, but the possibility that this was due to the difficulty of catching the tiny nymphs cannot be ruled out. Development of the greater part of the population was different from that described in that the nymphs remained between 2 and 4 mm long throughout the winter. The fact that a generation, with the possible exception mentioned above, remained within narrow size limits made the species a particularly good one for the study of life history (Fig. 7).

The small overwintering nymphs grew throughout their second summer, reached the last instar by the end of it or in the autumn, and emerged the following spring (Fig. 7) Dr J. H. Lawton finds that this is the usual life history in Durham. The 1957 year-class in Hodson's Tarn (see Pl. 3*a*) was exceptionally numerous. To begin with development was as usual, but in September 1958 the nymphs ranged in size from 3 to 11 mm with a large peak between 5 and 6 mm and a smaller one between 9 and 10 mm. The nymphs in the smaller size group, the more numerous one, did not grow during the winter but the others did and they emerged in the following year. The rest took a second

34

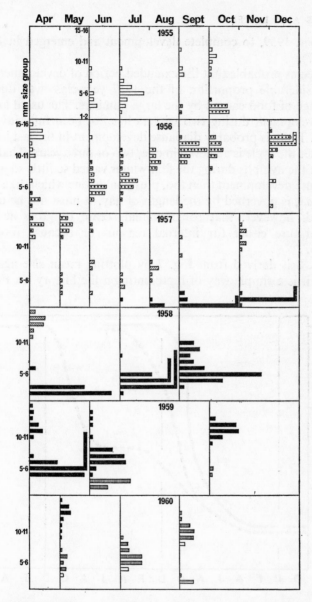

Fig. 7 *Pyrrhosoma* in Hodson's Tarn; numbers in each size group. Scale: a column equal to the width of one month represents twenty specimens. (After T. T. Macan, 1964. *Int. Rev. Hydrobiol.* **49**, p. 330.)

35

summer, 1959, to complete development and emerged in 1960 (Fig. 7).

It seems probable that the extended period of development of a considerable proportion of the 1957 year-class was due to shortage of food caused by the large numbers. The usual arrest of development during two winters has not been fully investigated. There is probably diapause development in the last instar whether the cycle is achieved in one, two or three years. That the size of the nymphs during the first winter varied so little suggests diapause development then too, presumably one which, like that of *Anax*, is governed by the length of day. It must not be overlooked, however, that some animals cease to grow at low temperature either for internal reasons or because food is scarce.

Fig. 8 is derived from Fig. 7 by plotting mean size against time; it is a simple way of representing a life history. In Fig. 7

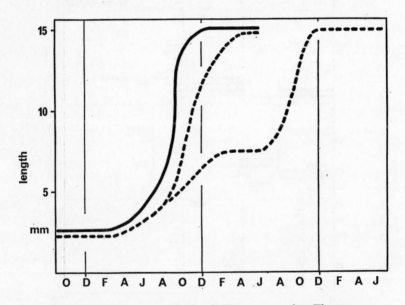

Fig. 8 *Pyrrhosoma* in Hodson's Tarn: mean size. The continuous line shows the development of a typical year-class, the broken line that of the 1957 year-class, which was exceptionally large.

the data are presented in full for any reader to interpret in another way if he sees fit. Dr Russell-Hunter's figures (Figs. 11–13) are similar but more elegant in that each histogram is plotted symmetrically about a central axis instead of from a base. It will be noted too that he measured to fractions of a millimetre. If the tops of the histograms in these figures had been joined, the result would have been a figure of the type seen in Fig. 6. The purist deplores this on the grounds that the line joining two measurements indicates numbers at intermediate sizes, which is not known. This is undeniable but the error must be small.

The main purpose of the above digression was to illustrate the ways in which life history can be regulated, but it also indicates that there is so much diversity that no rigid scheme of classification can be entertained. However, unless some pattern is to be discerned, description becomes an annotated catalogue and the wood is lost for the trees. The number of generations per year or years per generation is probably the best criterion on which to base any arrangement.

Most fish have a short spawning period. The new generation is ready to spawn after an interval which may be several years and which depends on size attained rather than on age. Once mature, specimens spawn each year until the end of their lives. *Arctocorisa germari* is one of the water bugs which can fly but which spend most of their lives in the water, and overwinter in the adult stage. Eggs are laid from May until mid-August, and then the adults die; a life span of several years with reproduction annually is rare among invertebrates. There are five instars, the first found from June to late July, the second from mid-June to early August and so on up to the fifth which occurs from early August well into October (Fig. 9). Thereafter the population consists entirely of the new generation of adults. This cycle of rapid growth in summer and a long period when only adults occur is typical of the beetles and the bugs, two groups of insects which are aquatic throughout life.

Lestes, whose life cycle has been described, is another example of a one-generation-a-year species, but it differs in that the winter is spent in the egg. *Ephemerella ignita*, an ephemeropteron to whose adults fishermen apply the name blue-winged olive,

37

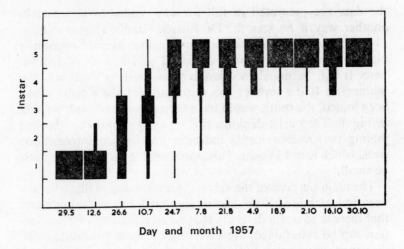

Fig. 9 Life history of *Arctocorisa germari*: percentage of nymphs in each instar on successive dates. (From D. T. Crisp, 1962, *Arch. Hydrobiol.* **58**, p. 265.)

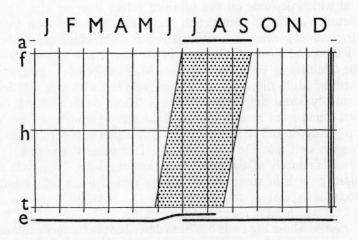

Fig. 10 Life history of *Emphemerella ignita*. a, adults; f, full-grown nymphs; h, half-grown nymphs; t, tiny nymphs; e, eggs. (From *Sci. Publs Freshwat. biol. Assoc.* No. 20.)

spends even longer in the egg. These hatch in June and complete development in about two months (Fig. 10). This, however, is not universally true and in the south of England a few adults may be found at any time of year. The life cycle of such specimens has not been elucidated. Some specimens of *Anax* also complete development in a year, but they overwinter as large nymphs. Yet another variation – growth in winter and dormancy in summer – has already been illustrated by reference to *Rhithrogena semicolorata* (Fig. 6). Incidentally it will be noted from Fig 6 that, although most eggs are laid within a month, very small larvae occur for some six months. This distribution could be due either to delayed hatching of some eggs or to a delay in the start of the growth of small larvae. Only in *Baetis rhodani*, another ephemeropteron, has it been shown definitely that some eggs take longer to hatch than others. It could be a method of avoiding the disastrous overcrowding that would ensue if all eggs hatched within a short period.

Many snails lay eggs in the spring or early summer and then die, but in this group flexibility is encountered once more, and the

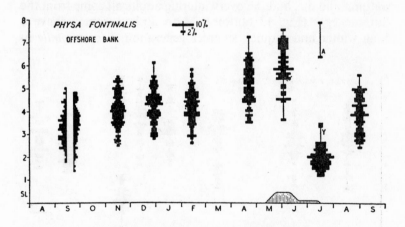

Fig. 11 Life history of *Physa fontinalis*: histograms show percentage in 0·1-mm size-groups in successive months and vertical shading shows number of eggs (A = adults, Y = young). (From W. Russell-Hunter, 1961, *Proc. zool. Soc. Lond.*, **137**, p. 146.)

snails introduce a third category, those that achieve two generations in a year. The shrewd observer will have noted that in this
account of life histories it has not been possible entirely to avoid
classification according to structure, which was the purpose of
introducing alternative systems. The snail, confined to the water
throughout its life, is in a stronger position to exploit what the
environment has to offer than such insects as the dragon-flies and
mayflies which must ensure that their adults are on the wing
during the optimal conditions of warm summer days. Dr W.
Russell-Hunter has recognized four types of life history, all of
which may occur in one species. Here two of them may be
coalesced. First there is the straightforward one generation a
year (Fig. 11). Secondly is the partial second generation when a
few snails, probably those that hatched earliest, lay eggs in late
summer. The overwintering generation consists predominantly
of snails that hatched in spring with a few that hatched in
autumn (Fig. 12). Third, when conditions are favourable, and
Dr Russell-Hunter is not prepared to state how far high temperature or good food supply are involved, come two complete
generations a year. All the adults hatched in spring spawn in the
autumn and die, and the overwintering adults all come from the
autumn eggs (Fig. 13 bottom). Many Ephemeroptera have a
long winter and a quick summer generation. *Cloeon simile* in

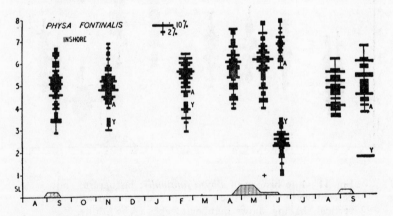

Fig. 12   Life history of *Physa fontinalis*: for explanation see
legend to Fig. 11. (From Russell-Hunter, *loc. cit.* p. 147.)

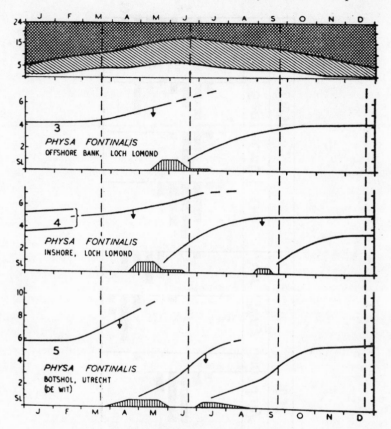

Fig. 13   Life history of *Physa fontinalis*; average size in successive months. At the top of each series are shown, in hours per day throughout the year, the potential period of sunlight (light shading), and the actual sunshine (white), the latter being based on mean figures for the Glasgow area in the years 1948–57. Vertical lines indicate number of eggs. (From Russell-Hunter, *loc. cit.* p. 160.)

Hodson's Tarn (Fig. 14) is a good example of this. In the same tarn *C. dipterum* achieves only one generation in a year (Fig. 15) though elsewhere it is known to achieve two. Since the adults of the summer generation are sometimes smaller than those of the overwintering generation, length is not the best guide to the

41

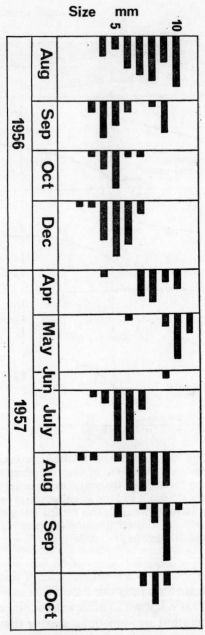

Fig. 14 Life history of *Cloeon simile* in Hodson's Tarn. Numbers in 1 mm size groups in successive months. Scale: distance between first two vertical lines = 10 specimens.

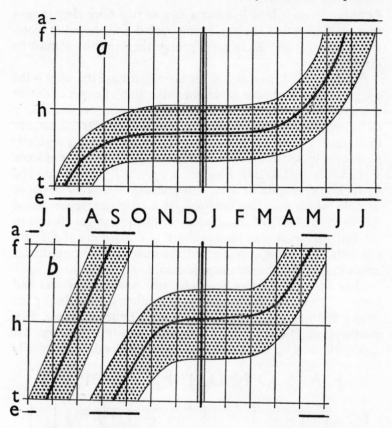

Fig. 15 Life histories of *Cloeon dipterum* (*a*) and *C. simile* (*b*) in Hodson's Tarn. Maximum, mean and minimum size in successive months. (From *Sci. Publs Freshwat. biol. Assoc.* No. 20) a, adults; f, full-grown nymphs; h, half-grown nymphs; t, tiny nymphs; e, eggs.

stage reached. It is useful to recognize a category of 'ripe' nymphs, these being distinguished by the dark wing cases and indeed, throughout development, length of the wing cases relative to body length indicates more than the latter alone. The interpretation of results is more certain if, in addition to measurements of the growing stage, the period during which adults are caught in an emergence trap (Pl. 1*a*, facing page 64) is known.

Ephemeroptera adults live but a day or two since they cannot feed, and therefore the egg-laying period and the adult-emergence period coincide. In other groups the oviposition must be observed.

Emergence is a fruitful field for investigation. First what is the trigger and second what determines duration? Length of day as a trigger has been discussed. Absolute temperature is unlikely to be important because it varies so much from year to year, but in Hodson's Tarn a rapid rise in temperature occurs regularly some time in April and there is evidence that emergence of some species is related to this. Emergence may be brought to an end by high temperature or some other change that kills larvae or nymphs which have not emerged. Is a long emergence period due to the inability of small larvae or nymphs to get enough to eat until the larger ones have emerged? Confinement of different numbers of specimens in cages of the same size offering similar conditions might answer this question.

Most Plecoptera (stoneflies) have one generation a year and none in Britain has more. The nymphs of most grow during the winter and aestivate as eggs, as do some Ephemeroptera. This phenomenon is commoner in running than in still water, but *Leptophlebia* (Fig. 16) is one example from ponds. Incidentally

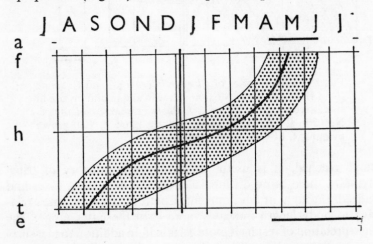

Fig. 16 Life history of *Leptophlebia* spp. (lettering and source as in Fig. 15).

this is a source of great disappointment to many students who plan a summer vacation's investigation of a stream fauna. Every year brings a few requests for advice to the present writer, who has to point out regretfully that so many species are present only as eggs in July and August that any collection is so incomplete as to be useless. There are a few species of stonefly which grow during the summer.

The stoneflies introduce another category, for some of the large carnivorous ones take two or three years to complete development.

At the other extreme come those which produce more than two generations a year. Most are Crustacea. Animals in the plankton, notably the Cladocera or waterfleas, reproduce rapidly by parthenogenesis when food is plentiful and the water is warm. When conditions begin to be less favourable males appear and the product of sexual reproduction is a resistant egg, in which stage the animals tide over the unfavourable period. This adaptation, as it appears to be, is noticed again in the next chapter. Of the larger Crustacea *Gammarus pulex*, the freshwater shrimp, has a long reproductive period. It starts near the beginning of the year, when the water is at its coldest and the days at their shortest, continues throughout the summer when the reverse obtains, and comes to an end in October, when a period of three months' rest begins. It poses fascinating physiological problems on which little work has been done. Dr H. B. N. Hynes, to whom we are indebted also for much of the information on stoneflies, has worked out that a female that was ripe in December could produce six broods before the end of the season, by which time her oldest daughters would be reproducing too. Shrimps are frequently seen in pairs.

The eggs of this shrimp are so large that they cannot pass down the oviduct in its usual state, for the terminal part is rigid as is the whole of the outer surface in the arthropods, to which the Crustacea belong. This means that increase in size is only possible after the animal has shed its hard external cuticle, which it does at intervals, and the new one is still soft. At this stage the eggs can pass down the oviduct and they are shed into a brood pouch between the thoracic legs. There the male pours sperm over them, which done he leaves the female and goes in search of another whose change of skin is imminent. The eggs take some three months to hatch at winter temperatures and 10–17 days in summer. The young shelter in the brood pouch for a few days and then disperse.

There is probably no time of year at which all the species abundant at some time in a pond are in the active stage and therefore anyone studying the composition of the community, the relation between predators and prey, or production must know about life histories. Anyone seeking to demonstrate the principle of the Eltonian Pyramid must take account of them, for the true shape of the pyramid can be ascertained only if calculations are based on the whole period of activity of the longest-living carnivore. A sample taken at one time in winter will show a very lean pyramid if the carnivores are dragon-flies that take two years to complete development and the main prey is a waterflea that completes many generations in a season.

It is common to find in scientific papers lists in which the species are grouped according to their structural affinities, Ephemeroptera, Plecoptera, Trichoptera for example. Ecologically a carnivorous caddis has more in common with a carnivorous stonefly than with a herbivorous caddis and it is suggested that a more informative grouping could be based on other categories as discussed above. Alternatively the order in which the species are arranged should illustrate some serial change.

Adaptation. Some freshwater animals came originally from the sea, and their main problem was physiological – how to maintain in the blood a concentration of salts different from that outside. A less permeable cuticle, a lower concentration in the body fluids, often coupled with a greater tolerance of fluctuation and an ability to take up salts from the medium or from urine before it is finally expelled are characteristic of freshwater animals in comparison with a marine relative. This field has attracted physiologists for many years and the advance of knowledge has been reviewed at intervals, most recently by Potts and Parry (1964).

The insects, some snails and some worms colonized fresh water from the land, to existence on which the insects particularly were well adapted, having an impermeable cuticle. This, and the concomitant ability to obtain salts from the food rather than from the medium, was a big advantage and facilitated the invasion of water in which the concentration of solutes was low. Some animals have changed little as a result of the transition, but others have adapted themselves to the peculiarities of the new medium in various ways. Methods of locomotion and respiration, for example, have been modified. In addition there have been adaptations that have enabled a species to occupy a particular niche. It is not easy to make a distinction between them and probably not profitable either. Here the policy is to draw attention to adaptations about which the reader can learn more by looking at a living animal than by looking at a book, and to describe less obvious ones more fully.

Some animals walk about on the bottom of a pond and among the plants in much the same way as their ancestors walked on the land. Others have modified some or all of their legs for swimming; many beetles and bugs, for example, have hind legs strongly reminiscent of the oars of a boat (Fig. 5). Feathering is achieved by segmentation and by hairs and bristles which collapse on the recovery stroke and remain rigid during the effective stroke. A similar arrangement on the cerci and caudal appendages together with undulations of the whole body enable some larvae to swim with great speed (Fig. 17). Dragon-fly larvae have

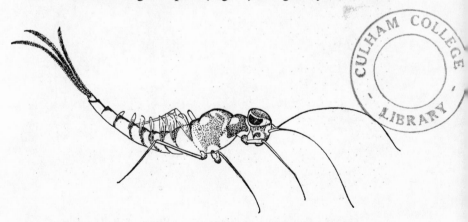

Fig. 17 Nymph of *Cloeon dipterum.* (Drawn by C. Joan Worthington.)

an anal respiratory chamber and can move forward rapidly by contracting this violently.

The immature stages of some insects obtain oxygen by diffusion through the cuticle. It will be recalled that the morphological school was content to name structures according to what their function appeared to be, an attitude which was regarded as uncritical by the generation that rose to prominence after the First World War. A well-known member of this latter school, the late Professor H. Munro Fox, was among the first to justify the new ideas, and his (Fox, 1921) demonstration is worth describing, for it is both simple and convincing. He found that grass taken from the garden behind the laboratory at Plymouth and steeped in water produced an infusion in which a protozoan, *Bodo sulcatus*, was numerous. This monad tended to congregate in areas in which oxygen was at a certain concentration. He took a larva of a chironomid, a family in which sausage-like gills on the last or last two segments are common, and placed it in his infusion under a coverslip. At first the flagellates were evenly distributed (Fig. 18a) but later they gathered round the larva as its consumption of oxygen produced the optimum concentration (Fig. 18b). As further respiration lowered the level near the body

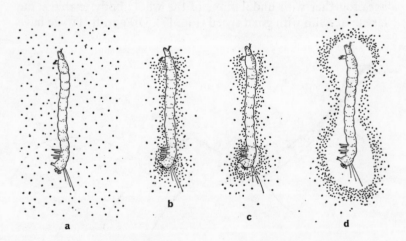

Fig. 18 Successive positions of a culture of *Bodo sulcatus* round a chironomid larva under a coverslip. (After H. M. Fox, 1921, *J. gen. Physiol.* 3.)

below the optimum, they moved away (Fig. 18c, *d*) leaving first those areas where exchange was most vigorous (Fig. 18c). It is clear from this figure that there was little exchange of gas between the water and the 'gills' on the penultimate segments. The greatest exchange in the figure is in the region of the thorax and at the hind end of the abdomen, but this was found to vary and gases passed through all parts of the body except the 'gills' mentioned. Later work with mosquito larvae showed that the function of these structures is the uptake of ions from the water.

Diffusion is a slow process, and many larvae undulate to remove from near the body the water in which they have reduced the oxygen tension. Measurement of the rate of undulation at different oxygen concentrations is another simple experiment.

Nymphs of Ephemeroptera often bear a series of thin plates down the sides of the abdomen. Work in Professor Munro Fox's laboratory showed that nymphs of *Cloeon* without gills respired as rapidly as normal larvae in water that was kept circulating. In still water they respired less rapidly. Normally these plates, which needless to say had been called gills, are in constant motion and their function is evidently to keep the general body surface bathed in a fresh layer of water.

Snails which have always been aquatic have protrusions from the body wall which are genuine gills. These were lost and replaced by a lung in those snails which colonized land. *Lymnaea peregra* is a freshwater species with terrestrial ancestry and Dr W. Russell-Hunter has shown that at temperatures below 12 °C it can satisfy its oxygen requirements by diffusion through the general body surface. In warmer water it must come to the surface periodically to refill its lung. This restricts it to shallow water. If a lake level rises after heavy rain, *L. peregra* may be forced on to stones, which, previously dry, are not coated with algae, and in consequence it goes hungry until algae re-establish themselves.

The respiratory system of an insect consists of tubes which ramify throughout the body, leading from the openings, which are known as spiracles, to the tissues where oxygen is required. Some aquatic larvae have retained this system, but generally all but the last pair of spiracles have disappeared, and these are

D

raised on a projection which pierces the water surface. The spiracles of culicine mosquito larvae are located at the tip of a conical 'siphon' (Fig. 19). Open spiracles make the possessor independent of dissolved oxygen and enable it to live in small productive pieces of water where the concentration may fluctuate greatly. Against this must be set the vulnerability inseparable from life at the surface, or from a life that involves periodic visits to the surface. Some larvae have solved both problems by developing a sharp termination to the region of the spiracles by which they pierce plants and establish contact with the air spaces. *Donacia* is an iridescent beetle often found above the water among the leaves of emergent plants. Its larva is a white grub that lives in the mud in which the plants are rooted and pierces the roots with two long black spines at the hind end (Fig. 20). *Mansonia* is a mosquito larva that pierces stems with its modified siphon (Fig. 21).

In a highly organic environment such as shallow pools into which a dung-heap drains, there is a copious food supply of a certain type but so little oxygen that few species can take advantage of it. Those that can may be extremely abundant, which is Thienemann's second law of biocenotics, as will be mentioned

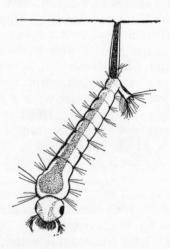

Fig. 19 A culicine larva.

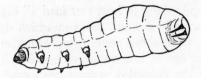

Fig. 20   Larva of *Donacia*, from the side and from below.
(Drawn by C. Joan Worthington.)

later. Anyone approaching a pool of the type described may see a forest of threads projecting up through the murk to reach the surface. A heavy step on the ground will cause them all to be withdrawn. If some of the filth is scooped up and filtered, the owner of the tail is found to be a greyish grub-like fly larva commonly known as the rat-tailed maggot. The spiracles are located at the end of the rat-tail which is telescopic and can be extended or contracted according to the depth of the water. The adult is a hover-fly. The correct name is *Tubipora*, though the earlier name *Eristalis* will be more familiar to older readers. Incidentally it may be asked where such creatures lived before

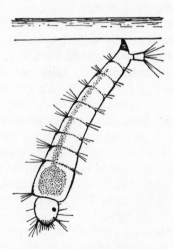

Fig. 21   Larva of *Mansonia*, attached to a stem. (Drawn by
C. Joan Worthington.)

51

man came to pile dung in heaps or lead all his sewage to one spot. Dead leaves in a small basin in which decomposition is rapid may produce these conditions, which are described as pollution when due to human activity, and it is in such places that larvae of the rat-tailed maggot are found too. The larva depicted in Fig. 22 was taken in an old quarry excavation so full

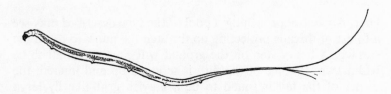

Fig. 22 A ptychopterid larva. (Drawn by C. Joan Worthington.)

of dead leaves that open water was confined to a small area in the middle. It is similar to the rat-tailed maggot but the tail is less retractile. It belongs to the family Ptychopteridae or Liriopidae and the adult resembles a daddy-long-legs.

Adult beetles and bugs have retained several pairs of spiracles, though these have generally shifted upwards or downwards so that they are in contact with a bubble of air carried between the back and the wings or trapped by hairs ventrally. The bubble is renewed at the surface periodically. When it is taken below the surface, it is subject to pressure and therefore it slowly passes into solution. However, since nitrogen dissolves more slowly than oxygen, the latter is passing inwards from the water to replace that used by the organism. The bubble is, therefore, functioning, as a temporary gill. If it could be enclosed by something rigid it would be a permanent gill. This has been achieved by means of what is known as a plastron. This is a pile of fine hairs each of which repels water. The result is that the pressure needed to force water into the pile is much greater than that experienced at depths at which the owner of the pile lives. Accordingly once its interstices are filled with gas, they remain filled and the animal is independent of the surface as long as there is plenty of oxygen in the water. This last is an important condition and plastron respiration is generally found in running water or on wave-beaten shores of lakes. Beetles of the family Elminthidae (Fig. 23) may be found walking on the bottom of streams slowly and

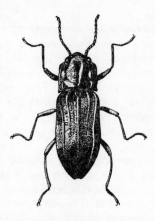

Fig. 23 Adult elminthid beetle (*Stenelmis canaliculata*). (Drawn by C. Joan Worthington.)

sedately as if they were land beetles. In contrast the bug *Aphelocheirus* has hind legs modified for swimming. It inhabits the deeper parts of rivers and is not often taken, probably because few collectors venture far enough into the water.

The retention of spiracles is imperative if an adult is to fly, and the power of flight must be of advantage to animals that live in small isolated pieces of water.

The main modification of feeding observed in freshwater animals is to take advantage of the small particles of dead plants and animals and small living animals and plants that float in abundance in productive water. This modification is found mainly in running water and others have been described already.

4 THE FIRST APPROACH

SUMMARY

A physiographical survey is the first step. Apart from temperature, information about climate from the nearest meteorological station should suffice. Temperature, transparency and level should be measured. Fluctuations in the concentration of oxygen are likely to be important, particularly in small productive ponds. The concentration of calcium should be known but few correlations with the concentration of the other common ions, except when they are far outside the average, have been uncovered by past work. Methods of sampling the animals and plants are described.

The surface dwellers may prove to be particularly suitable for studies of prey and production. The greatest variety of animals is found in the weeds and in the plankton, distinction between which is not clear in small ponds. In the mud are found groups in which the identification of species is difficult. The importance of identification to species is stressed. The food of carnivores may be determined from what is found inside them but herbivores and detritivores cannot digest everything that they eat. An investigation is suggested. Another problem is the frequency with which animals feed.

The community of a moorland fishpond is examined from the point of view of where and when each abundant species occurs and of how it feeds.

This chapter sets out to describe how to make a comprehensive survey and must start with the warning that this is beyond the scope of all but a team of full-time workers. An attempt is made, however, to write it in such a way that readers can select particular aspects which can be studied to some extent in isolation.

The first requirement is a map. The Ordnance Survey map will provide an outline that may be sufficient, but will be out of date if the pond is filling rapidly. If it is, a call at the local office of the Ordnance Survey may reveal that, although the published maps have not been revised for some time, an aerial survey has been made recently and the photographs may be purchased. If it turns out that the Ordnance Survey cannot provide an up-to-date map or a recent photograph, the task of surveying falls upon the naturalist. Edwards and Brown (1960) studying the vegetation of rivers, photographed them with a camera borne aloft on a home-made balloon. Should this prove beyond the capacity of the interested parties, there is nothing for it but triangulation. The first step is to insert conspicuous posts at intervals round the pond, posts which, if, left in position and numbered, later afford a convenient method of indicating the location of collecting places. The easiest way to define a triangle, in the author's experience, is to measure each side with a steel tape, an undertaking that is greatly facilitated by a cover of ice thick enough to bear. If the pond is too large for this, angles must be measured, and this is most accurately effected with a sextant. Less accurate is a prismatic compass. Alternatively a plane table may be used.

The contours will certainly have to be put in by the investigators. Here one may suppose that the Edwards–Brown balloon would be particularly valuable, assuming that the pond could be emptied. A photograph taken every time the level had dropped a known distance would show the contours with great accuracy and probably the vegetation as well. If an airborne camera is not available or the pond cannot be drained, the process will be more tedious. One method is to stretch a rope marked in suitable units across the pond and note where certain depths and changes in the vegetation occur. When the pond has been covered by transects in this way, the points are joined freehand.

Since the vegetation in a small pond may change considerably over a period of years or indeed in a short time, a careful map of the plants may be well worth making at the start.

Next comes the difficult decision of what physical and chemical variables should be measured. The records of rainfall and sunshine at the nearest meteorological station will probably be sufficient. The temperature of the water and of the air may not be correlated in a pond fed by an appreciable amount of ground water. Information such as that presented in Table 1 might prove valuable if available for different types of water in different parts

55

Table 1 Hodson's Tarn: number of hours at each temperature.
(Number of hours at say 7·5 °C is number above 7·0 and below 8·0 °C)

Temperature °C	1958	1959	1960	Hours 1961	1962	1963	1964
0·5	23	0	33	29	317	1195	179
1·5	198	10	617	486	552	529	563
2·5	691	896	530	449	591	149	903
3·5	1267	230	470	290	820	547	774
4·5	846	479	418	320	817	299	526
5·5	566	806	597	773	467	569	521
6·5	163	576	930	947	406	396	490
7·5	110	609	290	665	264	196	385
8·5	346	363	251	257	59	271	232
9·5	549	353	385	200	255	525	301
10·5	474	309	248	241	338	801	417
11·5	491	257	411	420	712	430	297
12·5	347	387	375	540	421	417	301
13·5	152	389	281	647	534	519	511
14·5	319	368	383	873	507	416	572
15·5	531	371	867	678	556	477	440
16·5	569	448	837	552	583	254	396
17·5	607	592	381	305	330	165	363
18·5	354	618	157	88	144	226	397
19·5	87	359	97		65	232	186
20·5	51	239	87		14	110	27
21·5	13	84	68		8	34	3
22·5	6	17	38			8	
23·5			12				
24·5			17				
25·5			4				

of the country. The figures quoted were obtained in a moorland fishpond in a rock basin in the Lake District by means of a recorder consisting of a steel bulb full of mercury connected to an arm through a capillary steel tube. The arm bore a pen which recorded on a circular chart moved by clockwork that required winding once a week. The one drawback to this method is that the recorder is difficult to conceal and is safe only in places to which the public has no access. The more modern thermistor (Mortimer and Moore, 1970) has been mentioned already. If continuous recording is not feasible, the next best thing is to conceal a maximum and minimum thermometer in the water, but this, of course, will not show how long the temperature remained at any given level. It may be advisable to place one such thermometer in shallow water and another inside a container in deep water. The container should hold an amount of water that will not rise appreciably in temperature during the short period that it is out of position for the resetting of the indicators. This process is carried out while the bulbs of the thermometer are still immersed. The purpose of the thermometer in deep water is to ascertain whether the pond is ever stratified.

The temperature of a pond may vary considerably during twenty-four hours and, in general, the smaller the volume of water the greater the range of temperature. Maximum and minimum values are, therefore, important, but the average may be the significant statistic in an ecological study. A recording thermometer is the only instrument from which all three may be obtained. There are various methods of measuring average temperature. An older one depended on the fact that the rate of inversion of glucose varies regularly with temperature. A more recent one makes use of the rate at which a substance passes through a capillary as its viscosity changes with temperature. This, the Hartley integrating thermometer is marketed by Linton Laboratories, Hadstock Road, Linton, Cambridge CB1 6HY.

The depth to which rooted plants can extend depends on the penetration of light, and light is cut off rapidly in many ponds frequented by animals. The opacity may be due to a thick growth of phytoplankton made possible by the enrichment of the water or to fine particles of clay kept in suspension by the paddling of the animals or to both. Various light-sensitive cells are available but it is generally sufficient to measure the depth at which a white disc, Secchi's disc, disappears from view.

The level of a small pond often fluctuates considerably and this may be a factor of such ecological importance that some kind of measurement is imperative if frequent visits are impossible. On one occasion the author wished to record the level reached during floods by a small stream that flows through his garden, and he put the question to his colleagues at lunch-time.

57

It provoked a lively flow of ideas till the end of the meal, by which time the picturesque stone-walled channel through which the water flows had been replaced by a regular concrete one, the well-tended lawn had been trenched to take electric cables, and the front room of the house had been filled with electric recording apparatus. It was highly instructive but not practicable. As the company dispersed, Professor Filteau, a visitor from Canada, mildly put forward the idea of a vertical board drilled at appropriate intervals with holes into which tubes could be inserted. The tubes slope downwards and therefore the number full of water indicates how high the water had reached at its maximum. This simple device has been in operation for many years. There seems no reason why it should not be used upside-down to record the level of a pond. The post would be attached to a flat base and submerged in the deepest part with all the tubes full. The lowest level reached would then be indicated by the position of the lowest empty tube. Anyone with a fondness for electrical apparatus could no doubt contrive a more complicated measuring device.

It will have been clear from what has already been written that, of the chemical factors, oxygen is likely to be of paramount importance. In productive ponds it may fluctuate violently from supersaturation during the day down to a low level, perhaps even zero, at night. Unfortunately the lowest level is likely to be reached just before it is light enough for photosynthesis to start, which at midsummer is very early in the morning, at a time when very few are prepared to be out taking samples. Moreover, it is difficult to forecast when a potentially catastrophic low level will be reached. Pennington has recorded that rotifers, or water-fleas, may attain such numbers that they eat the algal population till the survivors are too scarce to replace the oxygen which the animals, frantically active in search of food, reduce to a lethal level. The sudden death and decomposition of an algal population, which has possibly exhausted the supply of some nutrient, can cause deoxygenation. It may be imagined that a day so hot that a herd of cattle stand a long time in the water and enrich it could be followed by a night in which decomposition takes up all the oxygen. Any of these contingencies may not be disastrous unless the night is warm and still. It seems likely then that in some ponds a lethal concentration is attained only rarely, perhaps once or twice a season and the chances that spot analyses will detect it are small. Speculation will continue to surround the role of oxygen as an ecological factor in small ponds until continuous records are available. Mackereth (1964) has invented an excellent probe for measuring oxygen concentration and it can easily be attached to a recorder. It is available from The Lakes Instrument Co. Ltd, Oakland, Windermere, Westmorland but unfortunately at present it is expensive.

Substances released on deoxygenation react irreversibly with the tannin in certain woods to give a dark colour, and this has been put to practical use. Unfortunately the work (Dendy, 1965) was done in North America where the species are not the same as in Britain, but from the account it seems that oak is likely to be the best wood. Chestnut, sycamore and cherry are also possibilities. All conifers except larch are unsuitable. A freshly planed stake is driven into the pond bottom and the level where mud and water meet is carefully noted. After an exposure of less than one hour, the extent of the darkening will reveal the depth of deoxygenated water.

Deoxygenation is not necessarily a summer phenomenon and can take place under ice, particularly if the ice is covered with snow, a comparatively small depth of which cuts off all the light. In Canada it is known that ponds below a certain size are not worth stocking with fish because these will die of oxygen lack in the winter.

It is much harder to give advice about what other chemical analyses should be undertaken. Calcium is known to affect the distribution of various animals, though how far its action is direct and how far indirect is not known. There is probably no point in taking regular samples, and the extreme range may be discovered by making analyses towards the end of a wet spell and towards the end of a dry spell in both summer and winter. Fig. 24 shows some results obtained in this way by the author in an attempt to explain the distribution of snails in tarns in the Lake District.

Many analyses of the other common ions, sodium, magnesium, potassium, hydrogen, bicarbonate, chloride, sulphate and nitrate have been made, but they have not thrown a great deal of

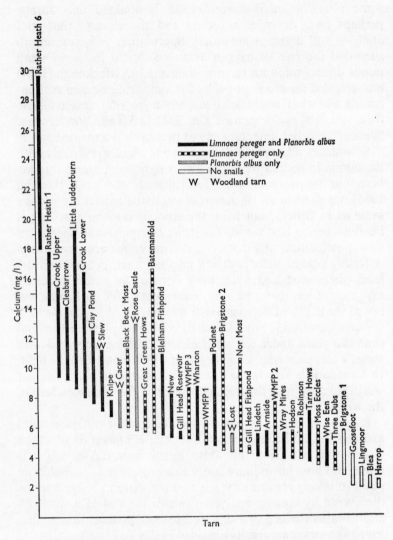

Fig. 24 Calcium concentration and the occurrence of snails in Lake District tarns. Each column extends from the minimum to the maximum concentration recorded; samples were taken towards the ends of wet spells and dry spells in both summer and winter. (From T. T. Macan, 1950. *J. Anim. Ecol.* **19**.)

light on the distribution of plants and animals, unless exceptionally high values are reached as they may be near the sea or in a few inland localities such as Droitwich. Phosphate and nitrate are two important nutrients which fluctuate with the wax and wane of plants, particularly algae, and which may limit production. In this, however, they may be less important than some of the so-called trace elements, substances which are present in quantities so small that their detection requires refined and specialized techniques. A great variety of organic compounds are present in water and when these can be identified as easily as elements, the picture may change radically. At present little is known about their identity.

Various metals such as copper, lead, zinc and aluminium are toxic when their concentrations exceed a certain level but this is rare except in polluted waters.

Mackereth (1963) has published a useful manual of water analysis in the Freshwater Biological Association's 'Scientific Publications' series. To conclude these remarks on physical and chemical factors, the author's suggestion, which is possibly not wholly without bias since he is a biologist, is to start on the animals and plants, to gain some idea of what factors appear to be influencing them, and then measure those factors.

Higher plants may be collected by means of a plant grab

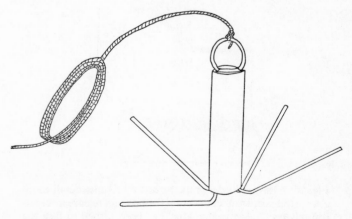

Fig. 25   Plant grab.

(Fig. 25) which is basically a lump of lead that can be pulled through the vegetation. Projecting forwards and outwards from this weight are three or four prongs of wire, of such strength that, if they catch a snag, a strong pull will bend them straight and free the grab. Plankton is caught in a conical net made of material, woven in a special way so that each hole retains its original dimensions. The material for a phytoplankton net should have forty-two strands to the centimetre, that for a zooplankton about half as many. The net may be designed to be fitted on to a pond-net pole, to be towed behind a boat or to be towed behind an otter board.

A sample of mud and its fauna may be collected in a variety of ways, but if sampling is to be quantitative a tube or box of square section should be used. The size depends on the density of the animals, and statisticians prefer a number of small samples to one large one. If a simple tube is driven far enough into the

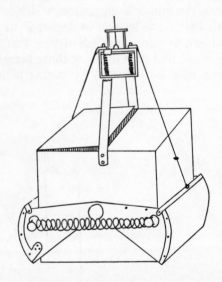

Fig. 26   Birge-Ekman grab. In order to show detail more clearly the near strap connecting the box and release mechanism has been cut and the latter has been turned to face the reader squarely.

mud and its upper end is then closed with a bung or a valve, a sample can be obtained with little trouble. Limnologists generally obtain samples from deep water by means of a Birge–Ekman grab (Fig. 26). This is a box, closed on top by two lids which are held vertical by the water on the way down and pressed flat by the water on the way up. The bottom is closed by two jaws operated by strong springs. Each jaw is held open by a bridle which clips on to a plate at the top of the box and through which the wire passes. With the jaws in this position, the grab is lowered until it sinks into the mud. A lump of lead is then sent sliding down the wire. It hits the plate, depresses it, and thereby releases the bridles. The springs close the jaws beneath the box and a sample is enclosed.

A collection of animals inhabiting weeds is commonly obtained by sweeping with a stout net, which, if pushed for a standard distance each time yields roughly quantitative samples. The net, however, turns out to be a surprisingly selective instrument; the proportion in it of animals such as leeches, which cling tight, is low and the proportion of animals which attempt to flee is high. A quantitative sampler that has proved satisfactory consists of two tubes fitting one within the other and armed with teeth at the lower end (Fig. 27). A boss on the inner tube is enclosed by a horizontal slit in the outer in a position which keeps the lower end of the two tubes at the same level. The inner tube, which is the longer, can be rotated to and fro within the limits set by the length of the slit. When this is done the teeth pass across each other in the same way as those of the machine used to cut a field of hay. It is desirable to operate the device from the stern of a boat to avoid damage to the vegetation by trampling. The lower tube is held in one hand and lowered vertically into the water. As it reaches the vegetation the other hand, gripping the top of the inner tube, rotates it backwards and forwards, so that the teeth cut the vegetation as they come to it, leaving within the tube a cylinder of vegetation whose height is the distance between the mud surface and the top of the plant being sampled and whose diameter is that of the inner tube. This appears to give a more accurate result than a grab which pushes the vegetation downwards and cuts it against the bottom on which it is growing. The cylinder sampler is driven

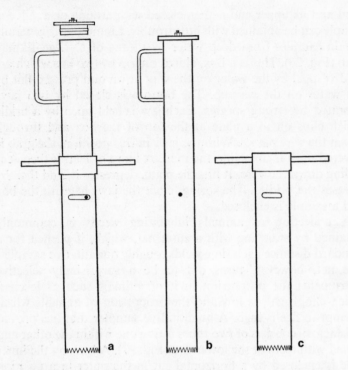

Fig. 27 Saw cylinder sampler (SCS). a, the sampler assembled; b, inner tube; c, outer tube. (From T. T. Macan, 1964, *Int. Rev. Hydrobiol.* **49.**)

for some distance into the mud to seal the bottom, and the top is closed by a lid consisting of two rigid discs with a rubber one between them. When the lid is in place, a butterfly nut is tightened to draw the rigid discs together, which compresses the rubber disc outwards against the wall of the tube. When the instrument is withdrawn any tendency of the core to stay behind causes a partial vacuum which helps to retain it. The sampler is emptied into a suitable container and the sample is poured into a large photographic dish. The plants and their roots are picked out and washed, which operation leaves the dish full of mud and water. The mud soon settles and extensive tests have shown that most animals can be separated from it by pouring the super-

(a)

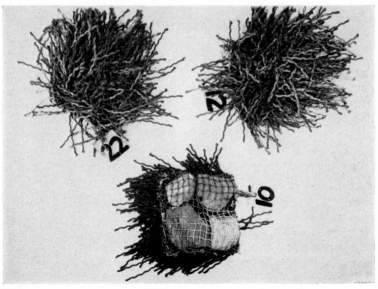

(b)

1a Trap for emerging insects. (*Photo*: T. Gledhill.)
  b Artificial *Littorella*. (*Photo*: E. Ramsbottom.)

2 Artificial *Carex*. (*Photo*: T. T. Macan.)

natant water through a fine net. Nearly all come out during the first two washings, though the standard practice is to fill the dish and filter the water four times. Some animals, such as leeches, snails and caddis larvae, tend to remain in the mud and they are secured by pouring the mud into a net, washing it thoroughly through the mesh, and sorting the debris left behind. All this takes time and at the end of it there are many fewer animals than might have been obtained with a sweep of a net. On the other hand the volume of vegetation from which they came is known.

Recently experiments have been carried out with small mats, 12.5×12.5 cm, of artificial vegetation. Artificial *Littorella* (Pl. lb, facing p. 64) has been made from polypropylene rope 13 mm in circumference. It is cut into appropriate lengths and un-ravelled, and the mat is made on a lattice of 'orange polythene shrimp netting' supplied by Gundry of Bridport in the same way as a wool rug. Each length is doubled and passed round one of the cross-pieces of the lattice. The free ends are passed through the loop so formed and pulled tight. A more supple net sown on to the bottom of the mat encloses stones to sink it. This tech-nique has several advantages. The animals can be dislodged by swilling the mat in water, and there is little debris from which they have to be separated. The smallest animals are caught because the mat can be brought up in a fine net inserted beneath it after it has been gently lifted from the bottom; a fine net used for sweeping clogs with algae rapidly, and a surprisingly high proportion of animals small enough to escape through the meshes of the conventional coarse net do so. The mat can be replaced and examined again after a week or two, for colonization appears to be rapid. Mats provided with bridles and floats can be brought up from depths at which other methods of sampling are difficult.

This technique appears to serve well if the problems are any of the following: elucidation of life histories, distribution in different parts of a water body, movement within the water body, or the relation of number of animals to density of vegetation. If, however, the object is to discover the absolute numbers, samples of the real vegetation must be taken as well, for preliminary observations indicate that there are more animals per unit area

E

in the artificial than in the real vegetation and the proportions of the various species in the two are not the same.

For a comparison of numbers and number of threads per unit area, artificial *Carex* (P. 2, facing p. 65) has been found to be more suitable. It was made of lengths of whole rope and a more rigid lattice of polyester mesh supplied by Henry Simon Ltd of Manchester. The lattice measured 25 × 25 cm and the lengths of rope were 45 cm long. They could be fused to the lattice with a touch of a hot metal rod. The point about polypropylene rope is that it floats, and therefore the strands in the artificial *Carex* rise vertically when the lattice to which they are attached is weighted and sunk.

Some of the animals collected will be nymphs or larvae that cannot be identified; the chironomids and caddisflies are the most important groups for which keys to the immature stages are not yet available because not all the species have been described. A reliable list of the species present can be obtained by breeding the adults from the larvae in some artificial system, but it is easier to trap the adults as they emerge. A number of traps have been devised and described. A simple one is a box that fits into a frame buoyed by floats (Pl. 1*a*). The top should be of plastic or some similar material to keep the rain out and some of the sides may be similar, but at least one should be of netting or small insects will be trapped in moisture that has condensed. A device that is useful for emptying one of the traps is illustrated in Fig. 28. Insects in the trap tend to try to escape towards the sky and can be picked out easily if the trap is kept tilted at the right angle.

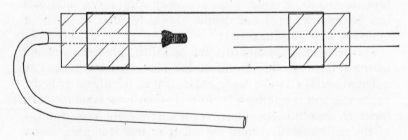

Fig. 28 Sucking tube.

The stunning of fish by means of an electric current is a method now widely used in flowing water, but it is not effective in standing water particularly if the water is clear. It is therefore necessary to use nets. The seine net is similar to a trawl but is hauled into the bank by parties on the ropes at either end, having previously been carried out in a boat and laid out in such a way that a suitable U-shaped area is encircled. One disadvantage of the seine is that it pulls up weeds, and, if it is desired to leave these undisturbed, gill nets must be used. These are simple curtains with a foot rope that sinks and a head rope that floats. They are now made of fine nylon thread and are available in a variety of mesh sizes.

At this stage a decision must be taken about what is to be investigated, for only a team of experienced workers could cope with the catches from all the apparatus described. The surface dwellers, to revert to the categories mentioned briefly on page 25, are not important in the general economy of a piece of water, and not much is known about them. For this reason, and because there are not many of them – twenty-three species of Heteroptera, twelve species of Coleoptera in the family Gyrinidae, and a few springtails – they may appeal to some workers. They have several advantages. In the first place they are easy to see and, since they are often confined within small areas, an estimate of total numbers based on a count of the population of a small known area should be reasonably accurate. Their feeding can be watched. Professor W. H. R. Lumsden once spent several days beside a boathouse on Windermere collecting the insects which, having fallen into the water, had been seized by a water strider. The resulting list of species was surprisingly long. They appear to have few enemies, and, though they would seem to be particularly vulnerable to fish that feed at the surface, they are rarely found inside such fish. They offer, therefore, an opportunity to study control of numbers. Dr R. Brinkhurst has brought forward evidence that large striders ignore smaller ones until a critical number per unit area is reached, after which they eat them, but the work was left at a stage from which it might well be taken further. On still sunny days many insects are on the wing and some of these, coming to grief in the water surface, provide food for the striders. On windy days animals may be

67

blown out of trees and bushes on to the water surface. It should be possible to count the casualties inside a wooden frame moored in an area inhabited by a colony of striders. If curtains projecting upwards 5 or 8 cm were attached to the sides, the predators would be excluded and an accurate assessment of the amount of food available to them might be possible. These animals may prove more suitable for a study of relation between prey and production than any which live in the water.

In a pond with much vegetation the distinction between true plankton and weed fauna is not clear, for there are a number of species of Cladocera which, although they are often taken swimming in the open water, feed mainly on what they find on the surface of plants. The attraction of the weed fauna is its great variety, and those interested in fish food will find most of their material in this community. The mud fauna is not recommended. It is comprised mainly of three groups, chironomid larvae, pea-mussels (*Pisidium*) and oligochaetes. There are an enormous number of species of chironomid and at the moment there is no complete key to the larvae. There are fifteen species of *Pisidium* and though there is a key to them, they are difficult to separate. The oligochaetes are another group in which identification of some of the species is difficult.

The previous paragraphs will have made it evident that the author believes that identification to species is necessary, a belief which some may challenge. To bring forward support for it, a digression into the realm of applied science will now be made, for applied scientists have no time for studies that are not essential. The story is taken from a book by L. W. Hackett (1937), a book which incidentally is written in a style that makes it a pleasure to read. After the First World War malaria spread throughout Eastern Europe till, to quote Hackett's words, 'the king of tropical diseases set foot in the Arctic circle'. It then died down faster than could be accounted for by any measures taken against it and eventually was to be found in areas, mostly near the coast, where it had always been endemic. It was thought at the time that only *Anopheles maculipennis* carried malaria in Europe. Why had malaria broken out in a region that was normally free of it and why had it disappeared? The question remained unanswered for some years, though not for want of

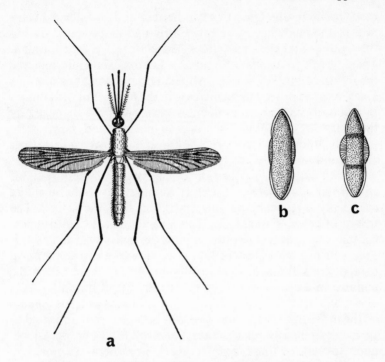

Fig. 29   a, *Anopheles maculipennis* ♀. b, Egg of *A. sacharovi* (this species has now been split from the *A. maculipennis* complex). c, Egg of *A. maculipennis typicus*. (T. T. Macan del).

attention to it by malariologists. Then a sanitary inspector named Falleroni noticed that the pattern on *Anopheles* eggs was not always the same (Fig. 29), and this set on foot a line of investigation which eventually solved the problem. Six different patterns on the eggs were distinguished and they came from six different 'races' of *A. maculipennis*; what exactly they were remained in dispute for some time. The races fell into two groups, one occurring inland, the other in slightly brackish water generally near the coast. Within each group the races had different geographical distributions. The important difference between the groups was that the inland one did not bite man if cattle were available, whereas the coastal one fed on man or on

69

cattle indifferently. Once this had been established, the explanation of the epidemic was clear. Cattle had disappeared during the fighting and the mosquitoes had perforce to feed on man. There were two ancillary favourable factors. The main one was the return of carriers, soldiers who had been serving in malarious areas, and the other was an increase in the number of mosquitoes in a countryside whose drainage system had been damaged by the war. As conditions returned to normal and farms were properly stocked once more, the mosquitoes reverted to their usual food and ceased to transmit malaria to man.

There remained the question of the status of these races of *Anopheles maculipennis*. Careful examination of the genitalia eventually revealed some differences, but they were small. The results of crossing were tried. This was not as easy as it sounds, for many species will not mate unless the males swarm, and this they will not do unless certain conditions, rarely achieved in captivity, are fulfilled. A few, in contrast, will mate within the confines of a test-tube. One of the 'races' was of this latter type, and it was, therefore, possible to cross its males with the females of the other five. When the cross was between what appeared to be the most closely related 'races', the result was an occasional adult, but all of them were sterile. A cross between more distantly related 'races' produced eggs that did not develop. There seem good grounds therefore for regarding the six 'races' as species.

This story illustrates more than the necessity for correct identification of species. It provides an example of the complex interaction between different kinds of organisms and of the trend which taxonomy is taking. At the beginning this is the domain of the worker who is generally to be found in museums, distinguishing species according to structural differences. But genetic difference is not necessarily reflected in structural difference, and it may be revealed by the work of the ecologist and the physiologist.

That identification to species takes time must be faced. The author's experience is that long consecutive periods at the beginning are desirable. The ideal situation is a field course where an expedition in the morning leaves the rest of the day available for identification. At the end of the first day a class is often dejected and perplexed, by the end of the second there has been enough success to arouse interest and fire ambition, and thereafter the class is surprised to find how greatly the difficulties have diminished. A wise teacher will, of course, steer students away from the most difficult groups, and also from unsatisfactory keys. To some groups there are no keys, as already mentioned, because the necessary work has not been done. To others there are keys that are incomplete, poorly illustrated or unsatisfactory for some other reason. During the last few decades keys to a number of groups have been issued by the Freshwater Biological Association and these may be purchased from the librarian at Ferry House, Ambleside, Westmorland. Those who cannot assign an animal to the group with confidence may find a book by the present writer useful; it assumes slight knowledge of the animal kingdom and takes the reader to a point from which other keys continue (Macan, 1959). Advantage is taken of each new impression to bring the references up-to-date.

There is one final question to be answered before the community can be viewed as a whole; what does each species eat? Most carnivores swallow large morsels of their prey and the contents of their digestive tract can be identified. It may be necessary to put a name to a single leg or a head but this is not difficult for anyone who is familiar with the animals available at different seasons. Fig. 30 shows the food of trout in Hodson's Tarn. *Eurycercus lamellatus* is one of the largest Cladocera, but even so it is a small morsel for any but a small trout. Insufficient is known about its natural fluctuations to make a statement about number eaten and number available. Throughout the summer all kinds of insects are flying about, and these are preyed on extensively by the trout, either when they come to grief in the surface film or when they approach it to lay eggs. During this time chironomids in all stages are eaten in large quantity; the pupae coming to the surface, and the adults just after emergence are obviously at risk, but it must be assumed that larvae become restless before pupation and expose themselves. During the rest of the year they are evidently secure in their tubes in the mud or on the surface of plants. As the summer passes, the plankton animals become scarce and few insects are on the wing. It is a

71

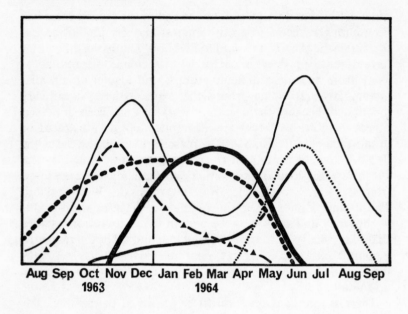

Aug Sep Oct Nov Dec Jan Feb Mar Apr May Jun Jul Aug Sep
1963 1964

scale		symbol
5	Enallagma	——————
5	Phryganea	▲— ▲— ▲
100	Eurycercus	————
50	Leptophlebia	▬▬▬▬▬
10	Lymnaea peregra	●●●●●●
50	airborne	··············

Fig. 30   Food of trout in Hodson's Tarn.

period of scarcity for the trout. As winter draws on, it starts to feed on *Lymnaea peregra, Leptophlebia* and *Phryganea*, a caddis. These have been present in some numbers for a month or two previously, but are evidently not taken by fish until they have attained a certain size. The numbers of *Phryganea* eaten fall off fairly soon because the number available drops. The population declined in the same way when there were no fish to eat them. The numbers of *Enallagma* eaten increased in early summer, and this increase is probably related to restlessness of the nymphs before emergence. The curve for *Pyrrhosoma*, not shown on

72

Fig. 30, is similar to that for *Enallagma* but reaches its peak some weeks earlier. The interval between the two is similar to the interval between the peaks of emergence.

This, it need hardly be stated, is a simplified diagram and at any time of year the diet of any one fish is varied. Occasionally an individual is found to have been specializing and it contains large numbers of one animal.

Bugs suck the body fluids of their prey and flatworms digest much of theirs externally. Dr T. B. Reynoldson has studied the food of the latter by means of a serological technique. An example of every possible species of prey is crushed and the body fluids are injected into a rabbit. In due course flatworms are collected and each one, having been crushed, is tested against a drop of blood from each rabbit. A cloudiness indicates that the flatworm had eaten the organism to which that particular rabbit had been sensitized. It need hardly be pointed out that this technique is possible only under special conditions.

Such tests as have been carried out indicate that carnivores digest and assimilate their food efficiently, being able to make use of almost everything but the hard parts. The feeding of herbivores and detritivores is less easy to study. It is not difficult to establish what they are eating, but an answer to the question of what they are digesting and assimilating is harder to obtain. Many algae pass unharmed through the alimentary tracts of animals, presumably protected by their cell walls. It is likely that few if any freshwater organisms can produce an enzyme that will break down the cell walls of plants that have died and contributed to the debris on the bottom of a pond. The nourishment of animals that feed on this material must therefore come from the fungi and bacteria which are attacking it and on the protozoa and other small animals for which it provides a substratum. Little is known about this microflora and fauna and information about it in the future may well go far to explaining the distribution of larger organisms.

A complete study can be made only in well-equipped laboratories by people trained to use special techniques, but there is no reason why simple experiments should not provide useful pointers. Some hardy animal that grows fairly rapidly – *Asellus* might prove suitable – could be reared indoors on a variety of diets and its rate of growth and reproduction followed. A sterile diet is out of the question, but a fresh batch of sterilized leaves every day would approximate closely to this, as bacteria and fungi would not have time to reach significant numbers. Comparison could be made with a culture provided each day with leaves seeped in water for a given length of time under conditions where attack by decomposers could be assured. Kaushik and Hynes (1971) have shown that some deciduous leaves, elm for example, decay more rapidly than others such as oak and that these are the leaves which are eaten most readily. Anyone concerned with ponds might need to test pine needles and *Sphagnum* fronds.

An important but difficult question is how often animals feed. How much an animal contained at the moment of death is easy to ascertain, and how long before an observed mass of food would have disappeared from the alimentary tract at any given temperature can be established. But how often does an animal eat its fill? The author has been surprised at the number of empty animals he has opened, and his colleague Dr J. M. Elliott, who has made extensive observations on stream animals at all times of day, tells him that this is because he works union hours and catches animals between 9 a.m. and 5 p.m. Collections made between 9 p.m. and 5 a.m. might be more informative; daytime is the dangerous period for small animals, when they are safer lying immobile and concealed where hunters cannot easily find them. This is clearly a line that should be followed, but will it explain the astonishing differences between growth rates of various animals? *Dytiscus* is one of the largest freshwater organisms but it is one of the quickest growers. It has been the subject of a very detailed study by Blunck (1924), whose work is worth special attention. It is one of those thorough studies that are commonly associated with the German race. It is so thorough that the reader feels that there will never be any need to repeat the observations. It is, however, in danger of being overlooked for three reasons; it was done long ago, it runs to a formidable number of pages and, for English readers, it is in a foreign language. There is a real risk that as the quantity of published work continues to increase, anything more than five years old will be regarded as out of date, and ground which has been covered

74

thoroughly will be gone over again, probably less effectively. *Dytiscus* requires 6 tadpoles in order to complete the first instar, but it generally eats up to twice this number if it can find them, and will consume as many as 30 if they are abundant. The second instar requires 18 and the third 130 tadpoles and the average and maximum consumption bears a relation to these figures similar to that in the first instar. If food supply and temperature are optimal, development can be completed in less than two months.

In contrast to *Dytiscus*, some of the large carnivorous stoneflies, which are also large as freshwater animals go, take three years to complete development. All our stoneflies grow slowly. They are also the animals that are abundant where food appears to be scarcest, at the top of mountain streams and in unproductive lakes. Have they a lower rate of metabolism than the animals that abound in more productive places, and if so can they make do with an occasional meal or do they feed continually on food of which they can utilize only a small fraction? Mann (1962) records that the medicinal leech can survive on one good meal every six months, but there is no information about the frequency of feeding of other species, some of which devour small animals whole. The best known example of fasting is provided by the salmon which appears not to feed for several months after it has entered fresh water and before it spawns. The metabolism of cold-blooded animals is a subject for experienced physiologists, but anyone with plenty of patience can make a contribution to the question of how often animals feed.

At last we are in a position to look at a community as a whole. Hodson's Tarn (Pl. 3*a*, facing p. 80) may be taken as an example. It is a typical artificial moorland fishpond about half a hectare in extent (1 acre) and about 3 metres deep at the deepest point. It lies in an area of unproductive soil much of which has been planted with spruce and larch. The drainage area is thirty times the area of the tarn and, if it be assumed that the average depth is 75 centimetres, $2\frac{1}{2}$ centimetres of rain is enough to replace all the water in the tarn, provided that the soil is saturated. This amount of rain in 24 hours is not infrequent in the Lake District and it is often accompanied by sufficient wind to mix the water in the tarn. It seems likely, therefore, that replacement of most of the water in the tarn happens several times a year, though no observations to confirm this and to discover the effects on the plankton have been made. In 1955 there were beds of beaked sedge (*Carex rostrata*) fringing the tarn on either side of the single inflow and floating pondweed (*Potamogeton natans*) between them. The shallow water in the rest of the tarn was covered with lakewort (*Littorella uniflora*) and in the middle there was a dense forest of milfoil (*Myriophyllum alternaeflorum*). During the next nine years this vegetation throve and the *Littorella*, particularly, increased in range, in size and in number per unit area. Then a decline started, the swards of *Littorella* began to retreat from the deeper water and the individuals became smaller and fewer. At the time of writing (1971) there is scarcely a plant of *Myriophyllum* to be found and, though *Potamogeton alpinus* has colonized some of the deeper parts of the tarn, large areas are bare. This condition is more typical of Lake District tarns than the earlier one of dense vegetation almost everywhere. It is an unusual phenomenon. No soil above water level remains bare long, provided it is watered sufficiently often. No explanation is known, but one of two possibilities is likely. Sterile patches in lakes are obviously connected in some way with organic matter derived from plant remains which have not decayed well. Similarly in Hodson's Tarn it is probable that, when the first observations were made, the plants were rooted in boulder clay. Now they are rooted in organic mud, derived both from their own remains and from dead leaves off the land. This mud, one must postulate, contains either a toxic substance or something that binds ions to it and renders them unavailable to the plants.

The first thing to discover about a community is its composition. As this varies with the season, the next thing to discover is the life history of the various constituents. The important thing is what each one does, which is far less easy to ascertain. The human peering down on a sward of *Littorella* through a glass-bottomed box is struck by his failure to see anything moving, although a net collection may have revealed a teeming animal life. Something might perhaps be observed in a narrow aquarium but the observer of an aquarium is nagged by two doubts: how much activity goes on after dark and how long do conditions in an aquarium resemble those in a natural body of water thousands of times greater in volume? What fun I thought to myself in an idle moment if, like Alice, one could eat something that would reduce one's size until one was half an inch tall and then wander through the forests of *Littorella* in the tarn. I thought again. Fun! It would be a nightmare to which only the pen of an Edgar Allan Poe could do justice.

However, let us suppose such an adventure to be possible and describe it, using such information about food as is available and some imagination. It will, however, be easier if, having shrunk from a height of 72 inches to half an inch, one returns to normal height and multiplies everything by 144 to retain the scale. In order not to allow fancy too free a rein, the numbers are based on those found in an artificial square of *Littorella* lifted on 21 September 1971. This square measured 12·5 × 12·5 cm and therefore our Gulliver, perhaps a better name for the adventurer than Alice, finds himself in a forest whose length and breadth is a little less than the length of a cricket pitch. There are twenty plants of *Littorella* or one to every 72 square feet, which means that roughly three of Gulliver's lengths lie between each one. In this way it is not greatly different from a terrestrial wood but here resemblance ceases. Each 'tree' consists of four or five trunks, 1–2 feet in diameter, tapering smoothly to a fine point between 20 and 60 feet above Gulliver's head. They do not stand up straight but diverge from the base, and, curving slightly, interlace to form a canopy. There are no leaves or branches but epiphytic algae break up the smooth outline and provide both food and cover for small animals. The forest is floored by the outer leaves which have fallen and which are in various stages of decay.

Many waterfleas sit in the canopy or swim from trunk to trunk and Gulliver compares them to the birds in the top of a terrestrial wood. He has no desire to climb up for a closer look, for through a gap in the canopy he has glimpsed what resembles a small airship gliding by. This is a trout and Gulliver can see it picking from the *Littorella* tips such waterfleas as have exposed themselves. Each is a tiny morsel for so large a predator and thousands are required to fill its capacious stomach. No wonder, thinks Gulliver, that its relative the salmon, which goes to the sea to feed on the much larger animals there, attains a size so much greater. No wonder, too, that the trout is on the look-out for larger pieces of food, and Gulliver decides that the deepest and thickest parts of the forest are the safest places for him. He watches the smaller herbivores carefully, for years of selection have no doubt produced a pattern of behaviour that keeps them out of the way of predators. *Leptophlebia* is the most numerous animal and there are just over a hundred in this small patch of forest. At this time of year they are not long out of the egg and measure but a foot from the front of the head to the end of the abdomen. They occur everywhere, on the leaves and on the bottom. Surprisingly scarce in this particular patch – there are only three – are larvae of chironomids, many of them living in the security of a secreted case either in the bottom or on the stems. They are of all sizes up to a length of about 6 feet and there are many species in the tarn, emerging at different periods during the summer. The dangerous time for them is when they must leave their cases to pupate and then travel to the surface to emerge. There are eighteen specimens of *Gammarus*, the smallest not much more than 1 foot long, the largest 6 feet. They move along on their sides and work their way underneath the debris, on which they feed, but they are catholic in their tastes and will eat any small animals they can catch.

Gulliver starts to explore cautiously. Fortunately for him his first view of the solitary full-grown *Pyrrhosoma* in his piece of forest is the tip of its legs gripping a leaf of *Littorella*. Carefully keeping out of range he works round to the other side where he can obtain a clear view of the monster, which is 10 feet long. It sits there immobile waiting for something to come within range of the labium, which is provided with two powerful teeth far enough apart to seize Gulliver round his slender middle and hold him while the robust mandibles and maxillae chew him into small pieces. There are five specimens of *Enallagma* but they all belong to the new generation and, a mere 1 foot long, are not large enough to attack Gulliver. The main food of these dragon-fly nymphs is chironomid larvae and waterfleas.

As Gulliver is watching the large *Pyrrhosoma*, a shadow passing overhead startles him. He looks up to see what might be a conger eel, twice as long as he is, swimming past, though

78

undulations are in the horizontal plane not the vertical plane as are those of fish. It is a leech. It is an alarming spectacle but, after watching it hunting, Gulliver realizes that it cannot attack him, for its circular mouth is not provided with the terrifying masticatory armature of the insects, and it cannot do much more than suck in its prey. It feeds largely, therefore, on sedentary organisms such as chironomid larvae and small snails. There are not many snails in Hodson's Tarn. Later Gulliver comes across a fragment of a *Carex* leaf, with the sides curving inwards like those of a trough. This is providing a protected resting-place for a second *Erpobdella*, though this one is not as secure as its relatives in the living *Carex*, lying in the sheath where a living leaf clasps the rest of the plant.

Gulliver need not fear leeches, nor dragon-flies as long as he keeps out of their way, and he begins to breathe more freely, but he keeps glancing in all directions none the less. Presently he sees a movement and hides behind the nearest stem to see what is coming. It is a hunter, *Phryganea*, provided with insect jaws capable of tackling prey which is too large for a leech and which does not often swim within reach of the lurking dragon-fly nymph. Its constant roaming makes it conspicuous but it is protected with a case of vegetable fragments glued together edge to edge in a neat spiral. The case is nearly 20 feet long and 3 feet in diameter and the larva crawls around with the immunity of a tank among riflemen. However, should it leave the cover of the forest a trout will snap it up case and all.

Gulliver realizes that he is too nimble for this enemy but envies the insects whose enormous eyes give a much wider field of vision than his own. He proceeds and comes to an open patch where the bottom is soft. He wonders whether he dares try to wade across and decides to wait to see whether anything moves on this apparently lifeless morass. Presently a *Gammarus* comes along on its side half crawling, half swimming. It starts to cross the open space when there is a flurry and it is struggling in the jaws of a *Sialis* larva that has been lying concealed in the soft fine debris. It is 10 feet long, and Gulliver compares it to a crocodile, speculating whether the broad horizontal gape of the jaws of *Sialis* is not more efficient than the vertical triangular one of the crocodile. It is fortunate perhaps that the structure of

79

the insect body prevents it growing large, for with six pairs of appendages from which to evolve a prey-catching and feeding apparatus it is more richly endowed than the vertebrate with its simple mouth.

Gulliver completes the exploration of his small patch of forest and encounters two larvae of *Limnephilus*, which look formidable in their untidy cases but which are herbivores. He feels grateful to the airship-like trout cruising overhead, for, were they absent, he might have encountered a *Dytiscus* water beetle twice as long as he or its even more formidable larva. Against them neither wariness nor nimbleness would have been any protection, and with this thought his adventure ends.

Gulliver encountered or thought about six types of carnivore, each exploiting the resources in a different way. There is a seventh, the net-spinning caddis larva, which is commoner in the *Myriophyllum* than in the *Littorella*. The main food is probably the waterfleas which blunder into the nets much as flies blunder into spiders' webs. The account of how each differs is based upon a limited amount of observation and some records of what was found inside each species. Any such account will, however, remain speculative in part until more direct observations have been made. There are tubes by means of which doctors and engineers can see inside opaque objects and perhaps these can be adapted for the observation of what goes on in weed-beds in fresh water.

Trout patrol the open water. *Sialis* is a typical inhabitant of mud and debris on the bottom. The success of the remaining carnivores depends on the structure of the vegetation in which they live. Observations on other types of vegetation may refute or confirm the observations that have been made. *Littorella* may prove to be one of the most suitable plants, for it is not too thick or too tall and, it does not die down in winter.

Ephemeroptera, Plecoptera, most Trichoptera, chironomid larvae, waterfleas and *Gammarus* are all herbivores but it is much less easy to be dogmatic about their food. They too rely on vegetation for cover, and their occurrence must depend to some extent on its nature.

The pattern of an aquatic community changes throughout the year. In Hodson's Tarn the only winter-growers are the two

(a)

(b)

3a Hodson's Tarn, an artificial moorland fish-pond. (*Photo*: J. Clegg.) *b* Pleated plastic square for sampling in midwater. (*Photo*: T. T. Macan.)

(a)

(b)

4a A cattle pond. (*Photo*: J. Clegg.)
b A farmyard pond. (*Photo*: J. Clegg.)

species of *Leptophlebia*. One may speculate on the selective advantage of the life history of this highly successful genus. Is it a cold-water species that cannot tolerate the temperatures of summer except in the egg stage? Or, being a cold-water species, does it gain advantage by growing at a time when most pre-dators are inactive? A third possibility is that its food is most abundant in winter. All the rest grow during summer, but larvae of the caddisflies are absent for a month or two during the development of the pupae and the eggs. Growth is rapid after the eggs have hatched and some larvae reach the last instar in a few weeks. Growth then ceases but the larvae of most species are active throughout the winter; only those of the swimming caddis, *Triaenodes*, hibernate. Comparison with the life history in warmer more productive ponds might be instructive.

In fifteen years the list of species in Hodson's Tarn has altered little, which is astonishing in a community that includes so many insects, capable of flight in the adult stage. There have been some changes in relative abundance, for which, it is hoped, an explanation will be found. The stability of a freshwater community probably decreases with diminishing volume of water, a line taken up again in Chapter 7. Once it is established that a community is stable, the question of what would change it may be posed. It may be possible to effect an experiment. Experiments on a large scale may be beyond the resources of even a research establishment, but they are often laid on by some human activity devoted to a different end. Windermere provides a good example. In the middle of its north basin three species of Ephemeroptera, three species of Plecoptera, and a net-spinning species of Trichoptera, all insects it will be noted, contribute substantially to the community on the stony shore in shallow water. At one of the regular collecting stations no flatworms have ever been taken, and *Asellus* is scarce at all of them. Farther north and farther south and throughout the south basin on the same kind of substratum three species of flatworm, *Asellus*, and the recently arrived amphipod, *Crangonyx pseudogracilis*, and the snail *Physa fontinalis* are numerous, and at many stations the insects mentioned have not been found (Fig. 31). The mem-bers of this second community are particularly abundant in the neighbourhood of the Bowness–Windermere sewage works, and

F

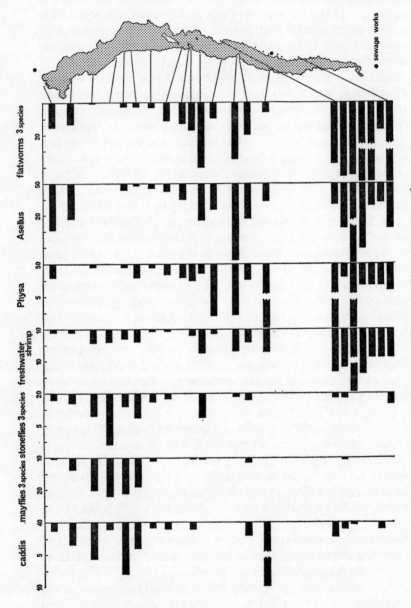

Fig. 31   Numbers at different stations on a stony substratum of some of the commoner animals in Windermere.

the conclusion that their numbers are associated with the enrichment of the lake water is irresistible. This sewage works, an
efficient one, probably affects the whole of the south basin up to
the point likely to be affected by the sewage from the numerous
cruisers moored among the islands that divide the lake. The
north end of the north basin is under the influence of the
Ambleside sewage works, which is also modern. Enrichment
from sewage effluent has resulted in a complete change of
community. Presumably it has somehow provided the food of
the flatworm – *Asellus* community, but nothing is known about
this, a gap in knowledge which illustrates the importance of a
study of the microscopic organisms in fresh water. It has been suggested that, when this food supply enables the flatworms and
other animals to become abundant, the insects disappear
because they are eaten. Their eggs perhaps are consumed by
detritivores.

If Hodson's Tarn were to be enriched there would no doubt
be change in the fauna, though not of the type described. Conclusions about less drastic changes can be based on comparison
with other tarns. For example the most numerous corixids in
Hodson's Tarn are *Sigara scotti, Hesperocorixa castanea*, and
Sigara distincta. None of these species is found above about 500
metres and at higher altitudes *Callicorixa wollastoni, Sigara nigrolineata* and *Arctocorisa carinata* take their place. In rich lowland
ponds the common species are *Corixa punctata* and *Sigara
limitata*, in brackish water *Sigara stagnalis* and *S. selecta*.
Various ecological factors that could bring about changes of this
kind are discussed in the next chapter. The communities on the
stony shores of Windermere suggest that comparisons are likely
to throw light on the factors governing composition, and comparisons that have been made or which might profitably be
made are discussed in Chapter 7. Chapter 6 is about how numbers are controlled. A community is made up of a number of
species about whose life histories, food, living-place and above
all numbers, something must be known. The consideration of
numbers leads on to production, the subject of the eighth and
last chapter.

5 ECOLOGICAL FACTORS

SUMMARY

According to the mutual exclusion principle two species with the same way of life cannot live in the same place. Some illustrations of the principle are quoted. Each species then has a distinct habitat and the reasons why it is confined to it are examined. Of the few species that have been studied extensively, Anopheles minimus *and two species of* Agrion *are restricted in occurrence by the behaviour of the adult, the ovipositing female and the territory – establishing male respectively. Other examples of behaviour are examined. Four species of flatworm live together in productive lakes because their food is not identical. In less productive lakes the food is more uniform and competition between the species leads to the survival of one only. Predation, particularly by dragon-fly nymphs, appears to be sufficient to keep small weedy ponds free of flatworms. Temperature may act directly, or indirectly through its effect on the rate of development of the species itself, on its food, its predators, or its competitors. Occasional deficiency of oxygen, also related to temperature may wipe out the population of small productive ponds.*

'Je variabler die Lebensbedingungen einer Lebenstätte, um so grosser die Artenzahl der zugehörigen Lebensgemeinschaften. Das ist das erstes Grundprinzip der Biozönotik' (Thienemann, 1954). Those who read German will note that this principle is phrased in simple terms, all the words but the last occurring in an ordinary dictionary. It may be translated 'The more variable the conditions in a living-place, the more species there are in the community inhabiting it. That is the first basic principle of biocoenotics.' The final word could equally well be translated 'ecology'.

'Principe de tendance a l'unité spécifique. Dans un milieu uniforme, restreint dans le temps et l'espace, ne tend à subsister qu'une espèce par genre' (Monard). In English: 'Principle of the tendency towards specific unity. There tends to be but one species per genus in a uniform environment in a given region at a given season. Later English versions run: 'Two species with similar ecology cannot live together in the same place' and 'complete competitors cannot co-exist'.

This sounds deep and abstruse, especially when quoted in a foreign language. A little reflection, however, reveals that these two principles are different ways of expressing the same idea, and that the idea is a simple one. This is true of many biological principles – once they have been formally expressed. The idea is a logical outcome of Darwin's concept of the origin of species by natural selection. If some genetic variation benefits its possessors in the struggle for existence, they are likely to survive, and they may give rise to a new species. This new species owes its success to its better adaptation to some particular mode of life, in which others cannot compete with it. In other words, each species has a peculiar habitat, and therefore within narrow environmental limits there is likely to be but one species per genus. The less narrow the environmental limits are, the more species there are likely to be per genus. The number of genera will be greater too; any naturalist would expect to find a more varied flora and fauna in a pond with varying depth and different soils in different parts than in a concrete reservoir with sheer sides and a flat bottom.

Monard's principle was overlooked by British ecologists who, for some years, attributed it to Gause. It gave rise to an enjoyable controversy. The opponents pointed out that it should work only when two species are competing for limited resources. If their numbers are kept below the level at which resources are insufficient by catastrophes such as floods, fires, high winds or drought, there is no reason why complete competitors should not co-exist. This is a fair argument but it probably applies only to populations living near the limits of their range. In water disasters of the kinds listed are much less frequent than on land. Even so the principle is difficult to prove. If two similar species are living together, the similarity may be no more than apparent and there exists some difference in way of life which the observer has overlooked. More often there is some cyclical change that favours first one species and then the other, and is revolving at a rate that never favours one species for long enough for it to replace the other. Thirdly there is always the possibility that one species is on its way to total victory but the process is slow in terms of human life span.

The value of a principle of this kind, particularly if it generates heat, is that it stimulates both sides to seek evidence to support their views. In the present instance it leads to a better knowledge of the habitat of each species and contributes therefore to a better knowledge about the community as a whole.

Hodson's Tarn provides some illustrations of how two species in the same genus co-exist without invalidating the principle. There are two species of *Leptophlebia, L. marginata* and *L. vespertina*, and in the Lake District it is unusual to find either by itself. However, *L. marginata* is larger than *L. vespertina* and emerges before it (Fig. 32). Unfortunately the two cannot be distinguished when small, but it is probable that the eggs of *L.*

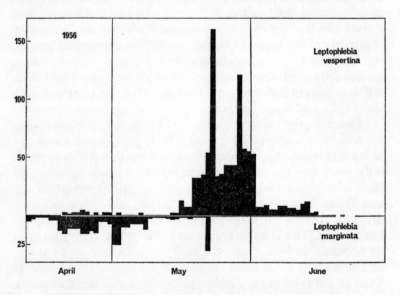

Fig. 32 Emergence of *Leptophlebia marginata* and *L. vespertina* (number of cast skins in frames on successive days). (From T. T. Macan, 1965, *Arch. Hydrobiol.* **61**, p. 278.)

marginata hatch earlier and that this species is, therefore, the larger throughout development. This difference in size could mean that the two were not eating exactly the same sort of food at the same time. Monard's formulation does not allow exactly for this situation but his reservation 'at the same season' indicates that he had envisaged the possibility of temporal separation. *Cloeon dipterum* and *C. simile* are also of different sizes at any given time, but there is a difference in place as well, for *C. dipterum* is more numerous in shallow, *C. simile* in deeper water. *C. simile* alone inhabits weed beds in lakes where wave action keeps the shallowest water free of vegetation. For the genus *Cloeon*, therefore, Hodson's Tarn provides a diversity which takes it outside Monard's concept of a uniform environment. The same is true of *Sigara scotti* and *Hesperocorixa castanea*, two species which, incidentally, alternate between being in the same genus and in different ones. This illustrates the point, important in this discussion, that the genus is a man-made category. No doubt it will be possible one day to base generic difference on genetic difference, when more is known about the latter. At present it is based on morphological difference with little regard to gene structure, physiology, or ecology, the potential importance of which was illustrated by the story of *Anopheles maculipennis* and malaria related above. However, to return to the two corixids, both are frequently found in one piece of water, but not in the same part of it, *H. castanea* inhabiting thick vegetation and *S. scotti* more open areas. Little is known about the reason for this. Two dragon-flies, *Pyrrhosoma nymphula* and *Enallagma cyathigerum*, though less closely related, are distributed in the same sort of way as the two corixids though there is more overlap (Fig. 33). Their ways of life appear to be identical, but their co-existence can be explained by the simple observation of the naturalist who happens to be watching on some sunny, not too windy, day when they are laying eggs. A pair of *Pyrrhosoma* remain attached during the process. They alight on a floating leaf of *Potamogeton natans* or an emergent leaf of *Carex* and the female lays her eggs in the leaf or the petiole, reaching down as far as she can into the water without wetting her wings. This confines the eggs to the shallower parts of the tarn, and most nymphs appear not to move far from their point

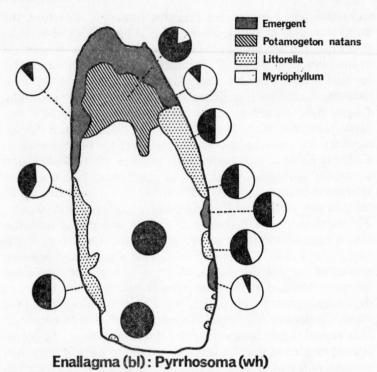

Emergent
Potamogeton natans
Littorella
Myriophyllum

Enallagma (bl) : Pyrrhosoma (wh)

Fig. 33 Relative proportions of *Enallagma* (black) and *Pyrrhosoma* (white) at twelve stations in Hodson's Tarn.

of origin. A pair of *Enallagma* fly low over the surface of the water in places where they can do so freely. They can be seen in the middle of the tarn where in the early years tiny inflorescences of *Myriophyllum* provided alighting places. From them the female walked down into the water, which caused the male to let go and fly off alone, and she might continue for a metre or so below the surface, periodically arching her abdomen to insert an egg into the plant tissue. After a period that might be as long as thirty minutes she released her hold and was shot rapidly to the surface by the bubble of air between her wings. Two species of *Deronectes*, a small water-beetle, occurred in the tarn when there were no fish, but only *D. duodecimpustulatus* survived their

88

reintroduction. Clearly there was some difference between them that had escaped human recognition.

It will be seen that consideration of Thienemann's first law and Monard's principle, now generally referred to as the competitive exclusion principle, has led to a study of the community by way of a study of its component parts. The problem is to define the habitat of each species and then to explain why it is confined to it. Some species, the snail *Lymnaea peregra* is a good example, are found in a wide range of conditions, others in a very narrow one; three species of mosquito, for instance, are hardly ever found in the larval stage except in water – filled cavities in trees. Macan (1963), in a book on the ecology of freshwater animals, devoted a chapter to each of the factors thought likely to be important. This approach is open to the criticism that, since the occurrence of any one species is the product of a number of interacting factors, the species, not single factors, should be the central points. This cannot be denied but unfortunately the distribution of few species has been studied in terms of all the factors operating on it, and the available information is about the temperature tolerance of species *A*, the oxygen requirements of species *B*, the behaviour of species *C*, the effect of predation on species *D* and so on. It is still impossible to avoid this treatment entirely, but the account which follows begins with species that have been studied from all the apparently relevant points of view.

In 1898 the Culicidae was an obscure family of the Diptera about which little was known. In that year Ronald Ross showed that mosquitoes carried malaria after which the family quickly became one of the most exhaustively studied in the animal kingdom. It was soon demonstrated that only the genus *Anopheles* transmits malaria and then that only a handful of species in it are important as vectors. Attention was turned to the breeding places so that control measures could be concentrated against the vectors and not dissipated in destroying harmless species. Field workers soon learned to recognize the breeding places and the systematists discovered differences in the eggs. It was discovered in the early years of the century that eggs transferred to places of a kind in which they never occurred naturally would hatch normally and give rise to larvae which developed through successive stages with no indication that conditions were unfavourable.

It was to contribute to an analysis of the habitat of *Anopheles minimus* that Dr R. C. Muirhead-Thomson went to Assam just before the Second World War. The species is a serious vector in the tea gardens and a fair amount was known about it. The larvae are almost always found in running water, never in small borrow-pits, which are generally cuboid excavations from which earth has been taken in such a way that the engineer can easily measure the volume which has been shifted. Larvae disappear when a small bush that casts dense shade is planted along the banks of a stream. Examination of the eggs established that there was no random oviposition and subsequent mortality of larvae that hatched in unfavourable conditions. Measurement of water temperature revealed that borrow-pits reached a maximum a few degrees higher than that reached in streams and, moreover, that at this temperature larvae of *A. minimus* died. This, however, was not as promising a line as might have been thought because it was soon evident that, in the small hours of the morning when the eggs were laid, water temperature was much lower. Study of the oviposition revealed two facts that were unexpected in view of the information available to Muirhead-Thomson when he arrived: the female sought shaded places in which to lay eggs and deposited them in stagnant water. As the investigation progressed it became clear that the absence of larvae under the bushes was not attributable to any deterrent influence by them on the female mosquito. What they did was to eliminate by means of their dense shade all other vegetation, with the result that there were no leaves of grasses and other plants trailing in the water. It was in the stagnant places under the lee of this overhanging vegetation that the mosquito found the conditions where she laid her eggs. Muirhead-Thomson then turned his attention to chemical factors and found that there was more organic matter in the water of the borrow-pits than in the flowing water. He suspected that it was different in kind too, but could not establish this in the light of the knowledge about organic compounds at the time. In fact he had reached an impasse that earlier workers on other species had also reached. Physical features played some part in deter-

90

mining the choice of the place to lay eggs, but ultimately the female was attracted by some particular organic compound giving off a distinctive odour.

It is unlikely that the next stage will be unravelled by anybody other than an experienced organic chemist. but some advance could come from simple experiments. *Culex pipiens*, a common British mosquito, breeds in a variety of places particularly domestic water-butts. It will breed in almost any container in which there is an infusion of organic matter. A simple experiment would be to offer it a variety of infusions, hay, dead leaves from different trees, flesh, garbage, to mention a few, or the same infusion at different ages. The containers should be in different places. Some tropical mosquitoes will enter houses to lay eggs, others will not. There seems to be no information about *C. pipiens* on this point, probably because organic infusions are not to be found inside Western houses. Another problem that may be mentioned here, though strictly it belongs to the next chapter, is whether females of this species will continue to lay eggs in a place where many have been laid already. The habit of *C. pipiens* of breeding in domestic containers opens up the possibility of experimenting with wild populations and the species has two further advantages: it lays its eggs glued together in the form of a raft which is easily seen and counted, and it does not bite man.

The *Anopheles minimus* story is not strictly appropriate to this section because the final clue is still missing, but it was worth telling for the way in which it illustrates the danger of the hasty conclusion in ecology. The explanation of the habitat of the next species, *Agrion virgo* and *A. splendens*, investigated by Zahner, is more complete. This beautiful dragon-fly with heavily pigmented wings may be seen flying near slowly-flowing water, in which its long-legged ungainly nymphs occur. The nymphs are strong clingers but in fast water they can only cling and cannot move about in search of prey. On the shore of a lake they do not react successfully to the advance and retreat of the waves and are dislodged by the alternating direction of the water movement. Zahner tested their oxygen requirements, confining them not in laboratory receptacles but in cages suspended in various natural bodies of water, the oxygen concentration of which was measured carefully. The outcome of this experiment was not clear cut, for, of nymphs from the same source incarcerated in the same cage, some would die soon, some survive till metamorphosis and some complete development. However, it was evident that the nymphs did require a concentration of oxygen higher than that likely to be maintained in still water. The nymphs are, on these findings, confined to slowly-flowing water by their oxygen requirements and inability to hunt in a fast current or to retain a foothold on a wave-washed shore. Study of the adults, however, revealed that the nymphs are found where they are because that is where the eggs are laid. The adults congregate beside some weedy slow-flowing stream and hawk up and down without interfering with each other. Over the water behaviour is different. Males establish territories each marked out by features of the vegetation, as far as the human can detect. From some central stem the male flies aggressively at any other male that invades his territory and generally drives him away though occasionally the invader is left in occupation after one of these clashes. A female is seized and pairing and egg-laying follow. It would seem that the distribution of the nymphs depends on the factors which make a piece of the water surface desirable territory in the eyes of the male.

Two British species in the genus *Agrion* were mentioned. *A. virgo* extends farther upstream than *A. splendens* to regions where the width is less, the current faster and the temperature lower.

The feature to be stressed in both the examples quoted is that the distribution of the aquatic stage is determined by the behaviour of the adult, and larvae of *Anopheles*, at least, would thrive in a range of conditions wider than that selected by the adult. There are many examples of this selection, particularly at the time of oviposition, but range of many species is determined by behaviour at other times as well. This is a fruitful field for exploration by those who have not access to elaborate equipment, for no more is needed than observation and simple experiment, unfashionable in recent times when the with-it scientist has grown accustomed to expensive apparatus. This brings us back to a point made earlier that work that has not been superseded by modern findings is always at risk of being overlooked because it is old. The purist would insist that any-body setting out to solve an ecological problem by observing what animals do in the wild must start at volume 1 of *The Zoological Record* and work his way through to the most recent number in order to make certain that the ground has not been covered already. However, those who are more concerned with gaining experience of solving a problem than with adding to knowledge will dispense with this time-consuming preliminary.

A few examples may be interpolated at this point. *Ancylus fluviatilis*, one of our two freshwater limpets, is one of the most tolerant of aquatic molluscs. It occurs commonly in swift streams and on lake shores; the author has found it in a small shallow quarry pool fouled by cattle and in a roadside horse-trough. The common feature of all these places, even, surprisingly, the horse trough, is a substratum of stones or bare rock. Other snails appear to thrive as well on a flat leaf as on a flat stone, but *A. fluviatilis* does not occur on flat leaves, though the other species, *A. lacustris*, does. *A. lacustris* is less tolerant than *A. fluviatilis* and is frequently absent from places where this species occurs on stones. Competition then is not the factor that keeps it away from leaves. One may put forward for testing the hypothesis that a specimen placed on a soft substratum, such as a water-lily leaf, will start to move and keep moving until it passes on to a hard surface. It should not be difficult to arrange a series of surfaces in an aquarium to test this hypothesis, but to the author's knowledge it has not been done. The reactions of other species might be investigated. Most of the snails found in Windermere are taken occasionally on stones but, apart from *Ancylus*, only *Physa fontinalis* is found on them sufficiently regularly to sug-gest that its occurrences are not due to casual wandering from nearby vegetation. Of the flat mayfly nymphs of the family Ecdyonuridae, only *Heptagenia fuscogrisea* is found on plants, though the surface of *Typha* and the large species of *Sparganium* appear to the human eye as suitable as that of a stone.

Different types of surface should not be difficult to test, nor should sediments that differ in size of particle. *Ephemera danica* for example is generally found in sand, *E. vulgata* in a finer more muddy substratum. Two species of American crayfish

93

offered two types of substratum in an experimental container tended to differ from each other in the one they burrowed into. Testing of the reactions of species which conceal themselves in debris or in vegetation will require the exercise of more ingenuity.

The digression may be continued to draw attention to other aspects of the suitability of *Ancylus* as an experimental animal. Its limited powers of locomotion make it more like a plant than an animal for the purposes of ecological study; or at least one supposes that, like the well-studied marine limpets, it does not move far. Macan (1970) suggests that its absence from Thirlmere is due to its inability to move up and down fast enough in that lake, where the rise and fall of the level since its conversion into a reservoir is much greater than in lakes that have not been interfered with. On stony regions of lakes and in streams, numbers tend to vary considerably from place to place, although conditions appear to be uniform. Little is known about its enemies, or indeed where any predator is able to pluck off a hard surface an animal that can cling tightly and offers little for a carnivore to get hold of. The eggs are likely to be more vulnerable and they pose the problem not only of what eats them but of what happens to the young on a thickly populated stone if numbers are not reduced by enemies.

To revert to the main theme, the last group whose occurrence has been explained in some detail is the flatworms, also inhabitants of the stones flooring exposed regions of lakes. Professor T. B. Reynoldson (1966) has made extensive collections in various parts of the British Isles, and finds that four species, *Polycelis nigra, Polycelis tenuis, Dugesia lugubris* and *Dendrocoelum lacteum* may occur together. Although they are not all in the same genus, all appear to have a similar mode of life. Places in which all four species occur tend to have a relatively high concentration of calcium and this is associated with high productivity. In Fig. 34 the lakes and ponds are grouped into six categories according to the concentration of calcium, and it will be noted that three species become progressively less numerous in the categories below the fifth. Two have been recorded once only in the second from lowest category and never in the lowest and a third occurs in all but in small numbers in the lowest. The bottom

94

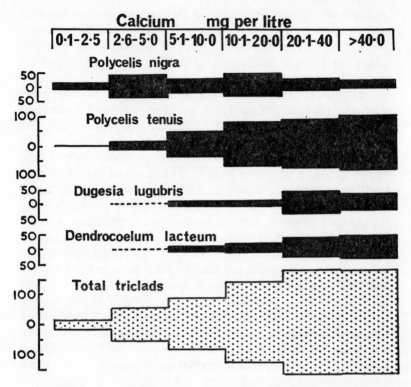

Fig. 34 The average abundance of the individual triclad species and the total population in lakes grouped according to the calcium content of the water. Dotted lines represent samples for a single lake. (After T. B. Reynoldson, 1966, *Advanc. Ecol. Res.* **3**.)

figure shows the average total number of flatworms and there is a marked decline from the fifth category downwards. In other words the more calcium the more flatworms. Reynoldson was able to remove most of the specimens of one species from a pond, whereupon those left behind multiplied until the total population was the same as it had been previously though the proportions of the two species was greatly different. Evidently numbers were limited by the available resources, and all the species were competing for those resources. There remained the

95

question of whether the absence of *Dugesia lugubris* and *Dendro-coelum lacteum* from waters with little calcium was due to competition, or to requirement by them of more calcium than the others for physiological reasons. This problem was solved by keeping all the species in cages in a small lake with a particularly low calcium content and in rearing them in the laboratory in water from the same place. It was established beyond doubt that none of the species was limited by calcium requirements over the range tested.

Reynoldson's method of investigating the food of these flat-worms has been described already. This revealed that there was a difference between the species. All feed on small animals, particularly oligochaetes, but *Dendrocoelum* eats more snails than the others and *Dugesia* more crustaceans, particularly *Asellus*. They are not therefore in direct competition. *Asellus*, however, is rarely found in waters with less than 5 mg/1 of calcium, and below this concentration the snail fauna is poor in both species and total numbers. In such waters, therefore, the four species do come into direct competition and *Dugesia* and *Dendrocoelum* are not successful. In a later paper Reynoldson and his school showed that the two species of *Polycelis* tend to feed on different oligochaetes and *P. tenuis* tends to be ousted by *P. nigra* at the lowest concentrations of calcium because the species that it feeds on more readily becomes scarce.

This is one of the clearest demonstrations of competition and of the competitive exclusion principle. The account attempts to follow a logical order not the chronological sequence of Reynoldson's ideas and findings. One piece of information that can now be seen as of basic importance was obtained during the course of the investigation when large numbers of one species were removed from a pond but total numbers were soon restored to the original level with the species that had not been collected contributing a much greater proportion. Evidence that total numbers were related to resources was also obtained in other ways. The flatworms happen to be uniquely suitable for an experiment of this kind. They breed throughout the warmer months but, when this leads to food shortage, developing eggs are resorbed and, if the supply remains inadequate, individuals decrease in size. As soon as there is a surplus, growth starts

96

again followed by reproduction. The adjustment of total bio-mass to available resources is therefore rapid. It is much slower in other animals with less flexible growth and reproduction, and competition between them would be harder to demonstrate.

A demonstration of this kind based mainly on field observations is more convincing than one based on work in an aquarium, where the possibility that results are affected by some unnatural condition can never be ruled out. Some, however, may wish to mount an experiment and they may find that *Asellus*, a hardy animal, is the most suitable for the purpose. Hynes and Williams (1965) have brought forward evidence that *Asellus aquaticus* is replacing *A. meridianus* in Britain. They followed up field work with culture experiments and found that, when the two were mixed, *A. meridianus* was generally eradicated. The classic work of this kind is that of Professor T. Park on flour beetles, a piece of research that has been going on for years.

The probable existence of competition can be recognized in the field if a place can be found where only one of two species occupying adjacent habitats occurs, and its range extends into the habitat occupied elsewhere by the other. Islands are the best place to look for this situation and Britain is particularly suitable for not only is the mainland an island but many smaller ones fringe it. For example the stonefly *Perlodes microcephala* has not been found in the Isle of Man and another large carnivore, *Diura bicaudata*, occurs from source to mouth in the streams there. On the mainland it occurs in the upper reaches and on stony lake shores, and the stretch between is occupied by *Perlodes microcephala*. In Britain *Gammarus pulex* may range from a spring high in the mountains down to the head of the estuary of a large river. On the continent there are two species unknown in Britain, one occurs in the headwaters and the other in the lower reaches; *G. pulex* is found only in the region between them. There is always a temptation to leap to the conclusion that any animal confined to high altitudes cannot tolerate the temperature prevailing lower down, but investigation has on occasion revealed that it is competition with some other species that limits its range.

The latest investigations of the Reynoldson school have been into the predators on flatworms. Here again they have used the serological techniques because the soft body of a planarian becomes unrecognizable so soon in a medium of digestive juices. Dragon-fly nymphs are found to take planarians readily and it seems likely that their numbers are sufficient in some heavily vegetated ponds to annihilate any flatworm population that does become established. Dragon-flies are scarce among stones and dead leaves, the substrata on which flatworms abound (Davies, 1969).

The three examples described above have illustrated behaviour and competition as ecological factors. Now some other important ones must be taken in isolation. The latest work of the Reynoldson school having introduced the topic of predation, it is logical to continue with an account of the experiment in Hodson's Tarn, to which several allusions have already been made. All the trout were removed and the rest of the fauna was studied for five years. Then the pond was overstocked with trout in their first year and the observations continued. At the same time samples of fish were netted at intervals in order that what they had eaten might be discovered. The overall result was no marked change in the community in the tarn but a shortening of the list of species, or, to be more accurate, the disappearance of some species from the regular collecting stations. Not all were annihilated, as specimens were occasionally seen or collected in places not sampled regularly. Tadpoles of both *Rana* and *Bufo* the large beetles *Dytiscus* and *Rantus*, and *Notonecta* were the animals most affected. It was not hard to see why. Tadpoles swim freely in all parts of the tarn with no attempt at concealment, the larvae of the water-beetles mentioned hunt in the open water, and *Notonecta* in all stages hangs from the water surface in wait for airborne prey that alights and is trapped by the tension. All these are easily seen by prowling fish, and it may be said that their behaviour is not adapted to co-existence with a large predator. Certain species which had been taken regularly when there were no fish but in small numbers were not taken again until the number of trout had been greatly reduced. At least they were not taken by the human collectors, possibly because the piscine collectors got there first, for a few were found inside fish. Possibly they were species whose typical habitat was not found in the tarn, and whose consequent restlessness exposed

98

them to the fish. It was the species that spent their lives in the cover provided by vegetation or debris on the bottom that were least affected though the range of some was curtailed. For example when there were no fish, *Hesperocorixa castanea* abounded in the shallow water of the tarn both in the *Carex* and in the thick luxuriant sward of *Littorella*. Evidently the latter plant did not afford sufficient protection from fish, for after they were introduced *H. castanea* was found only in the *Carex*, mainly in the shallowest water where *Sphagnum* provided additional cover (Fig. 35). During the first summer corixid nymphs were frequent in fish stomachs but in subsequent years they were very scarce. Some of the commonest species, *Leptophlebia, Enallagma, Pyrrhosoma* and *Lymnaea peregra* were almost as abundant as before although fish ate plenty of them, a phenomenon that is discussed more fully in the next chapter. It is suggested that they maintained reserves of small specimens that grew hardly at all but survived for a long time. The fish tended to eat only large specimens and presumably caught only those which had wandered too near the edge of the protective cover of vegetation. When this happened a good feeding place was left vacant for one of the small specimens to take over.

It can fairly be argued that all this is based on a *post hoc ergo propter hoc* argument and that the proof that change in numbers was due directly to predation by fish is not conclusive. It is unfortunately impossible to find two pieces of water so similar that one can serve as the control for an experiment in the other. It is also difficult to set up an indoor experiment in which the reduction in scale does not make comparison with events in larger bodies of water a doubtful proceeding. It is the sort of work where conclusions can be accepted with full confidence only when the experiment has been repeated several times and has yielded the same result. In the present instance some confirmation that the interpretation put forward is correct is provided by the return to former conditions as the population of trout was reduced.

The fluctuations in the population of *Lymnaea peregra* (Fig. 36) may be described to illustrate the danger of a too facile interpretation of the effects of predation. Between 1955 and 1960 the numbers of *Lymnaea peregra* did not fluctuate greatly and

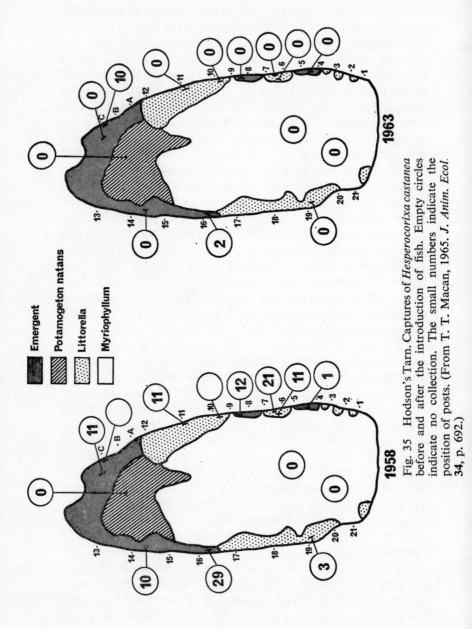

Fig. 35 Hodson's Tarn. Captures of *Hesperocorixa castanea* before and after the introduction of fish. Empty circles indicate no collection. The small numbers indicate the position of posts. (From T. T. Macan, 1965. *J. Anim. Ecol.* 34, p. 692.)

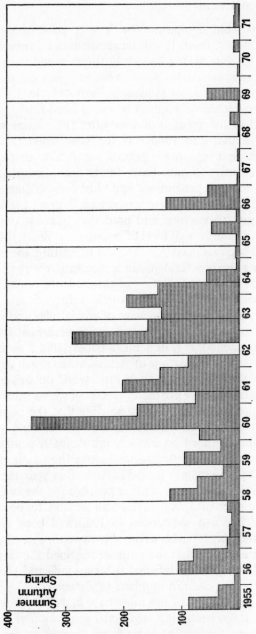

Fig. 36 Numbers of *Lymnaea peregra* at ten stations in Hodson's Tarn. (No collection in summer 1962)

maintained what comparison with later years shows to have been a moderate level. It will be recalled that *Lymnaea peregra* breeds in the late spring and early summer and then dies. The offspring may breed in the autumn but are often not mature until they are a year old. In Hodson's Tarn only in 1971 has unequivocal evidence of autumn breeding been found. Fish were introduced in the autumn of 1960 after the routine collections had been made, and 1960 was the year when the *Lymnaea* population first showed signs of significant increase. This, however, appears to have been due to exceptionally good survival of the new generation and therefore it started several months before the fish were introduced. During the time when fish were most numerous and predation heaviest, snails maintained an unusually high level of population. When the numbers of fish had been reduced substantially by netting, the population of *Lymnaea* crashed suddenly in a spectacular way (Fig. 36). In the absence of any explanation of these events, one may suggest possibilities that could prove worth testing. Snails are often infested with rediae and cercariae of trematodes and in extreme cases there appears to be more parasite than host tissue when the body is opened. It is difficult to believe that a snail supporting such a large population of parasites has much to spare for the production of eggs, though little work on this appears to have been done. There are many species of trematodes and many different alternative hosts. It is possible that the population of *Lymnaea* was kept at a moderate level in the early years by a parasite whose other host was a trout. Actually a small number of these survived the attempt to remove the whole population and the last was caught in 1957. After this year the infection would have died out and made it possible for the snails to produce more eggs and in a favourable season to become more abundant. The fish introduced in 1960 had been reared in a hatchery and may have been free of parasites. After a year or two it must be assumed that the parasite regained the tarn through the introduction of an infected fish or snail, and the resulting spread cause the drop in numbers of *Lymnaea*.

A large population of fish must have had effects beyond mere predation. It has been suggested that a generation of *Pyrrhosoma* or *Enallagma* consists of individuals occupying good feeding

102

points and growing well and others restricted to poor feeding places and growing slowly if at all (Fig. 7). Presumably when there were no fish and little mortality the starvelings, as they may be called, ultimately died. When there were fish and a steady removal of large dragon-fly nymphs, a certain number of starvelings could establish themselves in a good vantage point and grow normally. In other words predation led to greater production. Most of this was used by the fish as a source of energy and material for growth but some would pass to the pond bottom in their excreta. Moreover, fish took at the surface food that would not otherwise have passed into circulation in the water. Now Dr A. Pacaud culturing Cladocera found that they did not survive long in an infusion of lettuce leaf. If, however, he introduced *Planorbis corneus* to feed on the lettuce, the population lasted much longer. He obtained a similar result with cigarette paper, and surmises that the bacteria decomposing the snails' faeces provided suitable and sufficient food for the Cladocera whereas those decomposing vegetable tissue did not. It is possible that the fish faeces in the tarn affected the food of *Lymnaea peregra* in some way and made possible an increase in numbers.

There is wild speculation in both these hypotheses but they serve to illustrate certain lines which might profitably be followed.

The relation of predator and prey has been the subject of much theorizing and mathematical formulae which are beyond the scope of the present subject. Moreover, the field worker is likely to be faced with situations in which the variables are numerous, and change is a regular feature, whereas the mathematician must start with simple situations. Anyone interested will find a good account in *Animal Ecology, Aims and Methods* by Macfadyen (1957). Theoretically predators should increase in the presence of abundant prey and consequently reduce it. Many die of starvation, and the prey can increase once more. The numbers of the two continually oscillate out of phase. Macfadyen records that Gause, attempting to test this theory, never succeeded in obtaining more than a few oscillations before the populations died out, and suggests that the numbers were too small in the experimental conditions. In one of the most successful

103

experiments *Didinium* was the predator and *Paramecium* the prey. It was found that *Paramecium* would survive intense predation if cover in the form of sand grains was provided. It may surprise some readers to find that anybody could think of a predator–prey relationship in which cover was not a factor of supreme importance.

Another simple experiment described by Pacaud might be taken further and elaborated. He took four different predators and noted how many chironomid larvae they would eat in ten days at four different temperatures.

This leads to the next factor, temperature. It is unfortunate that the southern and eastern parts of England are not only warmer in summer than northern and western parts but generally lower, flatter, more calcareous and more productive. It is, therefore, difficult to disentangle the factors that confine some species to these two parts of the island. The collector who visits ponds and pools high up in the northern and western regions will take species that he will not find lower down and whose range may extend to Scandinavia, the Alps and possibly the Pyrenees, but little work has been done on their temperature requirements. Some are known to be tolerant of a wide range of temperature but intolerant of competition. Temperature tolerance is not difficult to study but anyone who wishes to work on animals from high altitudes must prepare a series of aquaria fitted with a cooling device and some means of transporting the animals from the mountain-top to the laboratory without a devastating change of temperature and other conditions. It is desirable too to know something of the environment from which the species comes. A peat-pool high up may be very warm during the afternoon of a still sunny day but at night much colder than a similar pool lower down. Some cold-water animals can tolerate high temperature for short periods.

The freshwater animals that are known to be restricted to cold water inhabit springs and the streams flowing from them.

At the lower end of the scale almost all waters in Britain reach the same temperature, a degree or two above freezing point and go no lower owing to water's peculiar property of being densest at 4 °C. The difference between a pond in the north and one in the south is therefore the duration of the period at this temperature. The exceptions mentioned are those waters which are warmed by the effluent from a cooling tower or some other industrial installation, or deliberately heated so that tropical plants may be cultivated. Various species have been able to establish themselves in these places. Certain species of bladder snail, *Physa*, about whose specific identity there is still no unanimity, have been in this country for some years. A recent and particularly spectacular arrival is the worm, *Branchiura sowerbyi*. Recorded from tanks for tropical plants at various places since late last century, it was first found in natural water by Dr K. H. Mann in 1958. He came across it in the River Thames near Reading below the warm effluent of a power station. Since then it has been found in the Bristol Avon and the canal connecting it with the River Kennet at two points not warmer than elsewhere.

These are examples where distribution is controlled by lethally high or lethally low temperatures, but within a restricted area such as Britain it is only rarely that temperature acts in this way. Generally more important is its effect on rate of growth, as was mentioned earlier when life histories were described. If temperature is high enough to enable a species to complete two generations a year, that species may be able to reproduce fast enough to make good losses from predators and parasites, whereas in a colder pond farther north one generation a year would not. Bacterial activity with its effect on decomposition as well as on total numbers increases as the temperature rises, and therefore in warmer water there is more food for larger animals.

Temperature and oxygen concentration cannot be separated. The higher the temperature the lower the concentration at which water is saturated with oxygen and the more rapid the bacterial activity which uses oxygen. That the fauna of rich lowland ponds is periodically wiped out by oxygen deficiency has already been suggested. The writer was recently collecting in some ponds of this type and expected to find *Asellus*, snails, leeches and other representatives of what some people call the permanent fauna. In fact the pond was inhabited by the temporary fauna, that is species which, though they may inhabit the pond permanently as long as conditions are favourable, can fly away when they are not, or which have a winged adult stage that can

105

recolonize a pond where conditions have been unfavourable. The ponds contained bugs and beetles, representatives of the first group and *Cloeon dipterum*, a well-known example of the second. A possible explanation of these findings is that the pond had been deficient in oxygen at some time in the past, but there are others; the author's diagnosis that it was a place suitable for *Asellus* may have been wrong. However, there can be no doubt that pond communities are annihilated by oxygen deficiency or that little investigation of the phenomenon has been made. The collection of definite evidence of it and observation of the subsequent recolonization is, therefore, recommended as an instructive study.

Oxygen leads on to other chemical factors, and it has already been stated that analysis of the common elements, apart from calcium, has not thrown much light on the distribution of animals and plants. Possibly analyses of trace elements and of organic compounds when they become easier or possible will be more illuminating. Calcium and good production go hand in hand, possibly because calcium promotes decomposition. Water with a low calcium concentration may be unsuitable to some animals for physiological reasons. There is much to be done in this field, but it is one for the experienced physiologist. Attention here is drawn to a few well-established facts. *Asellus* is rarely found in water with less than 5 mg/l of calcium; it may or may not occur when the concentration lies between 5 and 12·5 mg/l; it is generally found in places where there is more than 12·5 mg/l. For snails a concentration of 20 mg/l is critical and a number of species are hardly ever found in water softer than that. In Chapter 4 there was speculation about what species would be found in Hodson's Tarn if certain conditions changed, and corixids were mentioned. If its calcium concentration were raised to over 20 mg/l, it would contain four or five, possibly more species of snail in addition to the present one. The vegetation would of course be different too but there is no reason why conditions should be more variable. It would appear then that we would have here an exception to Thienemann's first law. The situation is, however, covered by his second law: 'The more the conditions in a biotope depart from the normal, and from the optimum for most species, the poorer in species and the richer in individuals will the biocoenosis be.' The classic example is the water polluted by sewage which provides a copious food supply for the few species that can tolerate the deficiency of oxygen. The second part of the law does not seem to be invariably true, for unfavourable conditions are not necessarily associated with a good food supply.

The paucity of snail species in Hodson's Tarn is therefore to be related to conditions optimum for most species. It is evident from field collections that the optimum is a high concentration of calcium, but whether this acts directly or indirectly through the food, as the enrichment of Windermere by purified sewage effluent apparently does, is not known. The conclusion is that the significant feature of a pond may be what is not present as much as what is.

6 CONTROL OF NUMBERS

SUMMARY
The numbers of a species are regulated by some factor inherent in the animal itself. This is commonly the establishment of territories. Tadpoles and snails produce a substance that retards the growth of small specimens. The Pacific salmon grows more slowly when numerous regardless of the amount of food available.

Animals with a short breeding season that attain maturity soon can provide a good food supply for a predator during a limited period only. A steady supply is provided by species which breed through most of the year or whose eggs do not all hatch at the same time. Also important as a sustained source of food are animals that set up a reserve of small specimens, from the ranks of which losses of large specimens can be made good. Life history is also discussed in relation to the origin of species.

This was the focal point of ecology a decade or two ago though the emphasis has now shifted to production. Perhaps one should start with an idea which probably owes more to theology than to biology but which is still prevalent in uninformed circles. It is the 'balance of nature', the vague idea being that animals are endowed with a reproductive capacity that provides herbivores with an increase in population sufficient to offset losses due to predators, and predators with one that balances losses but prevents numbers reaching a level where the prey is cropped to excess. A moment's reflection will reveal that this idea is not scientifically tenable. Animals preyed on must breed rapidly, their predators slowly, and any circumstance which favoured the prey could quickly lead to uncontrolled increase in numbers with consequent danger that all the food was eaten and a population died out over a wide area. There is a high *a priori* likelihood that there is some kind of self-imposed restriction on reproduction before catastrophic numbers are attained, though a substantial volume of evidence was not brought together until the book by Wynne-Edwards (1962) appeared.

One of the most common methods is the establishment of territories. This has already been noted in *Agrion*, described by Zahner, and it has been adumbrated that only a proportion of a generation of *Pyrrhosoma* or *Enallagma* nymphs gain a good vantage point for feeding, which is probably a territory. The most thoroughly attested example comes from running water, in which the habits of the trout (*Salmo trutta*) have been studied. Kalleberg in Sweden watched young specimens in a large tank with a plate-glass front. As soon as the fry had absorbed the yolk sac, each one established a territory which it defended with a ferocity that diminished with increasing distance from some central feature, generally a large stone. The size of the territory depended on the nature of the substratum; if it was rugged so that the occupier lurking under the lee of the what might be called its 'motte and bailey' could not see far, it was small; if the bottom was smoother the territory was larger. Fish unsuccessful in the scramble for territories were chivvied from place to place and eventually died. As the successful fish grew, they enlarged their territories, and there was a continual squeezing out of weaker specimens.

The author's colleague, Mr E. D. Le Cren, confirmed the Swedish findings in a field test. He divided a small stream into lengths of the same size and erected screens between each one, so that fish were confined to the area in which they were established. A different number of small fish were released in each area. The final population was, as expected, similar in each area and bore no relation to the original numbers.

In parenthesis it may be remarked that these observations have practical as well as theoretical importance. If the number of trout in a stream depends on the configuration of the bottom, the angler who contemplates stocking will be well advised to consider whether the stream is already carrying its maximum stock. It probably is unless there has been recent pollution, and the addition of more fish will achieve nothing. Moreover, both American and French workers have demonstrated that fish reared in a hatchery and supplied with excess and easily-available food every day are incapable of competing with wild fish that have been engaged in a fierce struggle for existence since the moment they used the last of the yolk with which their mother endowed them. It is perhaps not overfanciful to compare the wild fish to the followers of Genghis Khan or Tamerlane, trained to war from an early age, and the hatchery fish to the inhabitants of such places as Damascus and Baghdad who, after conquests in earlier generations, had adopted a more civilized and cultured way of life which led to their easy destruction.

The concept of territory as a means of keeping population in check has been familiar to students of higher animals for some years. Less is known about it among invertebrates, especially those that inhabit fresh water. The author is currently experimenting with populations kept in their home tarn in enclosures of fine netting and provided with artificial vegetation of the type already described. This is a simple type of experiment but one whose validity has still to be tested. A screen fine enough to retain the test animals, and bar entry to potential predators, must also keep out some of the animals that serve as food. Introduction adds a certain artificiality. Le Cren worked with the ideal experimental animal, one so large that a screen which kept it in did not keep its food out, and no predator big enough to attack it existed in the water. Herons and otters could be foiled by a screen which did not affect the aquatic animals in any way.

Incidentally Kalleberg noted that when the current in his aquarium ceased, or fell to a low rate, the trout left their territories and swam about in shoals. Little is known about control of populations in ponds and lakes, but there is relevant information from Canada, where the Pacific salmon has been studied with immense thoroughness. Those interested in this work and the topic generally will find an account of it in the *Mitteilungen der internationalen Vereinigung für Limnologie*, *13*, 1965, the proceedings of a symposium on 'Factors that regulate the size of natural populations in fresh water'.

The Pacific salmon lays its eggs in streams, but the fry drop downstream and develop in lakes. If the spawning beds are limited in extent, latecomers excavate the redds of early arrivals and expose their eggs to predators, and this acts as a control on numbers. Generally, however, there is enough suitable ground for every fish to make a redd without disturbing eggs already laid. Control then takes place in the lake. It was observed that the more fish there were per unit volume, the smaller was the average size. The obvious conclusion, not always the right one in ecology, was that this was due to shortage of food. Investigation, however, showed that the average size of salmon was much the same in relation to number per unit volume in lakes where plankton was plentiful and in lakes where it was scarce. Evidently crowding retarded growth regardless of food supply. When the fish grew slowly predators took a heavier toll of them.

The nature of the inhibition is unknown, but it has been demonstrated that tadpoles secrete a substance that arrests growth of other tadpoles. The larger the animal the more the secretion, and the smaller it is the more it is affected by the secretion. The result is that, if a number of tadpoles are kept together, a few large ones continue to grow and the smaller ones do not. This happens in water that has been inhabited by tadpoles, but not if the water has been heated. The substance, whatever it is, is evidently destroyed by heat. Inhibition of the same kind is practised by snails.

Dr M. Brown, rearing trout in tanks, observed that a lone fish did not use food as efficiently as one with a few companions and that when these exceeded a certain number efficiency diminished again. Possibly one is entering here the field of psychology. It has been known for long that mammals, rodents especially, exhibit signs of stress when they are crowded, and various abnormalities of behaviour, such as eating newborn young, check the rate of reproduction, but the mass psychology of lower animals is a field of which the exploration has only just begun.

The docketing of the various aspects of a phenomenon of this kind may be an instructive exercise for whoever does it but there is a risk that the product is something which, easily learnt by heart for the purpose of impressing examiners or colleagues, militates against original thought. The first steps generally involve large animals, which commonly means vertebrates, and the principles which emerge do not always take account of the immense versatility of simpler animals. The three methods of

111

population control described have all involved death of some of the population, but this is not universal. Flatworms, for example, do not die when overcrowded unless there is a very serious shortage of food. The reaction to this situation is first to stop breeding and then to grow smaller. Cladocera reproduce as fast as they can as long as conditions are favourable, abundance of food being one favourable condition. When conditions become adverse they produce resistant eggs which lie dormant until conditions are favourable once again. The advantages of attempting to study an entire community is that problems of this kind are seen in a framework and not in isolation.

Any discussion of numbers leads on to the practical question of what animals at the second stage of the food chain, or the second trophic level to use the current jargon, correspond to grass at the first? More simply what animals provide a source of food for predators throughout the year or for a considerable part of it? One thousand years ago man started to cut down the woodlands that covered the Lake District fells almost up to the summits to make pasturage for sheep, and the process eventually denuded most of the area. It is now covered mainly by grass, which can stand being constantly grazed by sheep because it grows from the base, not from the tip as do the dicotyledons. How can an animal respond to continuous grazing? As was seen in Chapter 3 many species have a single generation a year and a short growing period. The result is an abundance of food for a predator at the beginning of the generation and subsequently a steadily diminishing supply. The predator must hunt elsewhere. There will be a few species that are breeding when most are not, and the changing pattern of a carnivore's diet has already been illustrated by the trout in Hodson's Tarn (Fig. 30). There are, however, some species which provide a more continuous source of food. Mention has already been made of the Cladocera, which produce many generations in a short time while conditions remain favourable. Another Crustacean with a comparable life history is *Gammarus*, which starts breeding at the beginning of the year and produces young continuously until October. Only during the last three months of the year is there no breeding, which incidentally poses fascinating problems for the physiologist. Certain species, most as far as is known dwelling in streams, achieve the same effect by delayed hatching of the eggs. The oviposition season may be comparatively short but the eggs hatch over a long period and the population is continuously reinforced by new recruits. Also comparable to grass are those animals which maintain the reserves mentioned already. Observations suggest that a proportion of a generation of *Pyrrhosoma* or *Enallagma* does not grow because the nymphs cannot obtain a good vantage point for feeding. They survive, however, and form a reserve from the ranks of which any loss arising from the capture of a large specimen by a fish can be made good. There is, therefore, a supply of large nymphs available to fish over a period that is much longer than the oviposition and hatching season. This could possibly be regarded as an adaptation for life in the same piece of water as a large carnivore.

Another aspect of these different types of life history provides material for speculation. A contrast has just been drawn between two types of life history. Corixids provide a typical example of one where breeding does not last long, growth is quick, and for much of the year the population is represented by adults whose number must decline steadily. In the other numbers are maintained at a relatively high level over a long period by a long breeding season, by a long hatching period, or by the provision of reserves. All the animals discussed lay a large number of eggs and therefore a small breeding population will suffice to keep the species in existence. However, a small breeding population may mean that the habitat of the species within a body of water is not fully occupied and is open to colonization by another species, perhaps one with a slightly different optimum. It could happen perhaps that under these circumstances the small breeding population of one species would tend to congregate in one part of the habitat while those of the newcomer congregated in another. The outcome would be that what had been a wide habitat occupied by one species became two narrower ones occupied by two species. It might be expected, therefore, that, in groups with a corixid-type life history, species would be numerous and the habitat of each one narrow and sharply defined. This is certainly true of the corixids themselves. It is even possible that species might have originated in this way but this is highly controversial ground. One school maintains that a species splits into two only when the ancestral population becomes divided by a physical barrier. The main challengers are those who have studied large and old lakes such as Baikal and Ohrid. Both contain a high proportion of endemic species, often closely related, and it is difficult to envisage conditions in which the ancestors were geographically separated. The second school therefore maintains that differences in behaviour, in selection of a place to live, or in breeding time led to the evolution of species within one body of water.

It will be recollected that the surplus population of the trout perishes and there is no reserve of starvelings as in some other animals mentioned. The trout keeps its habitat fully occupied in a different way, by having no fixed size at sexual maturity. The size of the territory is not related to food supply. If this is poor growth is slow and the fish may never attain a weight of half a pound, if it is good a weight of several pounds may be reached. Reproductive powers are not impaired by small size though of course a small female contains fewer eggs than a large one. Food supply therefore determines total biomass independent of numbers.

7 COMPARISONS AND EXPERIMENTS

SUMMARY

Experiments in laboratory containers have thrown light on optimum space, numbers and food of such animals as snails, waterfleas, and Hydra, and have been used to investigate the effects of cropping and competition, though it is not always easy to relate the results to field conditions. The extent to which the composition of a natural commumity depends on chance is examined. The dispersal of plants, deduced from the study of ponds of known and different ages is slow. That of insects is likely to be faster and the impression is that this is true of other groups also, though the snails are the only one for which direct evidence can be brought forward. Studies of the dispersal of insects have been made by means of light traps and artificial containers, but neither give a complete picture since some species fly by day or are not attracted by light or by small areas of water. Details of the colonization in four years of a water tank that was emptied each winter are given. This leads to a discussion of the relation between volume of water and stability of the living community and possible experiments are outlined. Other comparisons discussed are those between places subject to different degrees of pollution; between places that dry up for varying lengths of time in summer; and between the communities in different types of vegetation.

Various activities that fall within the competence of those with limited time, experience and facilities have been described in the foregoing chapters. This one is devoted entirely to the topic. Some understanding of the relationships within a community can often be obtained through a comparison of two places, frequently two which differ mainly because of something that man has done. Laboratory experiments may also throw light on particular problems, and these may be taken first as the subject has already been broached in the last chapter.

The late Captain C. Diver made extensive investigation of

inheritance in snails and, in the course of it, made a number of observations on culturing them, observations which could profitably be elaborated. He prepared conditioned water by leaving a sprig of *Elodea canadensis* in water in a jar for three weeks, at the end of which period it was suitable for snails. One introduced into a 2-lb jam jar would grow larger than a similar one in a 1-lb jar. The number of eggs laid was also affected by the volume of living space, and it could be increased by frequent changes of water. Both growth and number of eggs were also affected by the number of snails when more than one was kept in the same container.

Water fleas can be cultured easily. Dr G. Fryer feeds his on a culture of the small alga, *Chlorella*, which is itself cultured in Chu's medium. This is made up from six stock solutions: $Ca(NO_3)_2$ 40 g/l, K_2HPO_4 5 g/l, $MgSO_47H_2O$ 25 g/l, Na_2CO_3 20 g/l, Na_2SiO_3 25 g/l. One ml from each of these stocks and 0–25 ml of normal HCl is added to one litre of water and boiled. When the liquid is cool, the medium is ready. Fryer obtains faster growth by shaking a little baker's yeast into a suspension in water and pouring it into the culture of waterfleas (see also Galtsoff *et al.* 1959 for culture methods).

Fryer's own work on the relation between feeding and structure calls for a skill in the manipulation of small animals that few will aspire to, but certain ideas put forward by Pacaud could be investigated without the necessity to master a difficult technique. He found that different species subsist on or thrive on different kinds of food and that this influences distribution. Bacteria, algae with and without a cuticle, and fine leaf fragments are some of the substances which he mentions. D. M. Pratt investigated the effect of temperature on a population of *Daphnia magna* and found that a temperature of 25 °C was more favourable than one of 18 °C, when considered in terms of rate of growth and reproduction. However, at the higher temperature reproduction was so rapid at the start that an unfavourably large population had been produced before excessive numbers began to check the birth rate, and the population died out. At 18 °C overcrowding depressed the rate of reproduction before numbers had reached a disastrous level, and the culture persisted much longer. From the point of view of survival, therefore, a temperature below the optimum for growth and reproduction was optimal. It is, however, difficult to apply results of this kind to field populations where food supply, predation, and a constantly renewed medium are some of the factors interacting to affect numbers of any species.

L. B. Slobodkin (1962) experimented with *Daphnia*, and with *Hydra* fed on the young of the brine shrimp, *Artemia salina*, which are supplied for the purpose by aquarists. The eggs hatch readily in brine solution, after which the resulting nauplii are washed in fresh water and then transferred to tubes which are immersed in ice-water. This immobilizes the nauplii, which sink to the bottom of the tube whence a suitable quantity can be transferred by means of a pipette to the bowls containing *Hydra*. Slobodkin was interested in the straightforward relationship between amount of prey eaten and amount of weight put on by the predator, in the effect of predation, and in competition. 'Predation' in these experiments was the removal of a certain number of specimens by the operator, and possibly 'cropping' might be a better word, for it is when man crops a natural population that he wants to know the parameters that Slobodkin investigated. Obviously the crop taken must not be so large that the result is a decline in the permanent population or stock, and on the other hand cropping which does not reduce a population to near this level is inefficient. Again it must be stressed that there is a wide gap between the findings of a simple experiment and practice in the field where so many more factors are operating. Competition between *Hydra littoralis* and *Chlorohydra viridissima*, which contains symbiotic algae, was studied, and the not unexpected finding was that the green *Hydra* is the survivor in a culture in the light. When the two are kept together in the dark, *H. oligactis* alone survives unless both are cropped heavily, when both survive.

It may be remarked that in order to obtain results of scientific importance from work of this kind the operator must spend a considerable amount of time caring for the cultures, changing the medium and preparing the food.

Comparisons in the field raise at once the question of how far any differences found may be due to chance. The first arrivals will be species that disperse readily and abound in some nearby

117

piece of water. It is likely that many are replaced as time goes on by species that are adapted to life in the conditions provided, but not much is known about how quickly or how completely. It may be remarked that, in general, workers who have studied one group come to recognize the habitat of each species, though they can rarely analyse the features of which it is constituted, and generally seem to find a species where they expect it. The impression is, therefore, that any piece of water is soon occupied by the species to which it offers a suitable habitat, but it should be stressed that this is an impression and not a statement based on hard scientific fact.

We now introduce at the same time the first of the comparisons and a point of view that contradicts the one just set forth. In 1923 H. (now Sir Harry) Godwin investigated the flora of seven ponds lying in gravel in the flood plain of the River Trent. One was an oxbow and the rest were pits that had been dug to provide ballast for a railway. All were about the same size and it seemed unlikely that there were marked physico-chemical differences between them. The point was that all were of different age, and Godwin was able to discover the exact year when each pit had been excavated. The oxbow was older and had been isolated about 1700. It contained forty-four species. The most recent, dating from 1902, contained eight species and the rest, with one exception, fell neatly in between into a series in which the numbers of species increased with increasing age. Only one species occurred in all seven ponds. Godwin draws no dogmatic conclusions but suggests that these findings indicate that the plants have poor powers of dispersal and that therefore the species found do depend to some extent on chance.

The author suggested that Thienemann's first law might apply here. Conditions in a recently excavated ballast pit are uniform and, as the years pass, the angles of the sides become less steep, humans coming to fish or cattle to drink will create little deltas and marginal depressions, which with other events, produce more variable conditions. Godwin agreed that this might play some part but maintained that the chief reason for the great number of species in the older ponds was slow dispersal. Though he did not say so, he was in fact maintaining that it was Thienemann's third law not his first that was being upheld. It runs: 'The longer conditions have remained unchanged, and the longer a place has offered an unaltered environment, the richer and more stable is the community in the place.'

Another study, that of corixids in dew ponds by Macan and Macfadyen, illustrates the sort of results that can be obtained by a more superficial study, much of the collecting having been done by Macfadyen while he was still at school. The factor here is not time since creation of the pond but time since it was last cleaned out. There was no direct information about this and it had to be judged from the condition of the dewpond at the time of the visit. Dewponds were seen at various stages and the past history of any one had to be guessed from what had been seen in others. A new dewpond has a concrete or clay bottom generally protected against penetration by hooves with a covering of rubble. As time goes by plants start to grow in a dewpond but, if it is an important supply of water to the farmer, he removes the vegetation frequently. Some dewponds are fenced off and the water is siphoned to a drinking trough with a ball tap; others are left open and the sheep or cattle can wade into them. In Fig. 37 the dewponds are arranged in order according to the population of corixids. In the top two, A2 and B40, only *C. punctata* was found. In the rest down to D2 the percentage of this species decreases steadily. From B10 down to the bottom the percentage of *S. nigrolineata* increases regularly. Several dewponds have been omitted from the histogram. All were heavily fouled by beasts and contained large numbers of *S. lateralis* and no other species. Evidently this is the only one that can tolerate these highly enriched conditions. Moderate enrichment favours all four species. As far as can be made out, a cessation of enrichment leads to conditions where the pond becomes overgrown with *Lemna* and in such places only *C. punctata* is found. The exclusive occurrence of *S. nigrolineata* is associated with the growth of rooted vegetation and the gradual disappearance of the pond as an aquatic habitat.

The classic study of transport of an aquatic animal other than an insect was made by the late Professor Boycott, who observed for many years the snails in a large number of ponds of which many had no inlet or outlet. On the average he found that a species turned up where it had not been found before

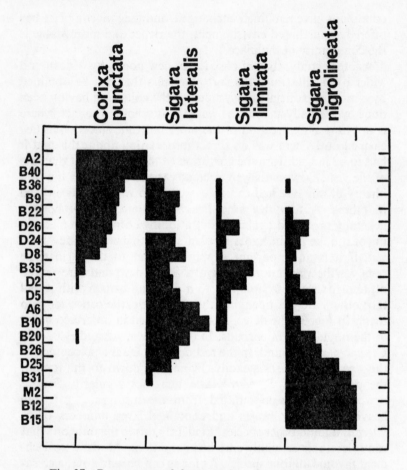

Fig. 37 Percentages of four species of corixid in twenty dewponds. (After T. T. Macan and A. Macfadyen, 1941. *J. Anim. Ecol.* **10**, p. 180.)

every nine years. This indicates a rapid interchange when allowance is made for the fact that a new arrival could be detected only when it belonged to a species not already present. Boycott also tried the experiment of transporting to ponds the large ramshorn, *Planorbis corneus*, which was not known in any of them but which was abundant in a nearby reservoir. It estab-

120

lished itself in some of the ponds, and the conclusion reached was that its powers of dispersal are poor.

Boycott's approach to the problem will not commend itself to any except the few who can contemplate long residence in the same area. Godwin's, on the other hand, could profitably be used by the young and transitory. If another such series of ponds could be found, and they cannot be rare in the flat parts of the country, a study of the distribution of even one animal such as *Asellus* would be worth undertaking.

Flight of insects has been investigated in recent years by means of the mercury-vapour light trap but unfortunately there are two interpretations of the scarcity of any species: it may not fly readily or, and this has been established for some species, it is not attracted to light or flies by day. Another approach has been to lay out tubs or other receptacles and see what flies into them. Results seem to suggest that the species that inhabit small ponds are more often on the wing than those that occur in larger bodies of water, which *a priori* is likely, but here too there can be more than one interpretation of the findings. Since beetles and bugs will also land on a glass plate, it seems likely that water from the air is to them no more than a shining surface, but it is possible that the species found in large pieces of water are not attracted to small areas of shiny surface. On the whole these lines of investigation do not seem very profitable.

The author obtained some unexpected results when invited in December 1938 to advise on what to do about corixids that infested a large concrete storage reservoir half a mile square. A customer was complaining that the bodies of corixids were clogging the screen with which he had protected his intake. The treatment incidentally was to put a screen over the exit from the reservoir, a measure which scarcely required professional advice from an entomologist. Large collections were made in December and in January, April and May the following year, and the unexpected aspect of them was that the most abundant species was different on each occasion and that a total of sixteen species was taken. Generally the species of corixids found do not change much with time and it is unusual to find more than about eight species. Extensive collecting has yielded only eleven in Windermere which offers a variety of different conditions; the reservoir with

121

its almost sheer sides and flat bottom was uniform. The explanation tentatively put forward was that this expanse of water was attractive to corixids flying over it, probably some of them in swarms, because it was so large. It turned out to be unsuitable and the insects eventually moved on. The large number of species was attributed to the absence of a permanent population, the presence of which, it is assumed, causes a more prompt departure, for other places rich in species had in common the likelihood of recent catastrophe that could have eradicated or driven out an established population. Here is another field in which useful comparisons could be made. In the Lake District there are a number of tarns all with a similar corixid fauna, and nine species or more were found only in four: two new tarns, one that was occasionally drained, and one in which black-headed gulls breed. The diverse fauna of the last was thought to be due to variable conditions, the pond being enriched by the birds during the breeding season and then, leached by the rain, reverting to a typical moorland type during the rest of the year. A pond in Scotland with nine species was near the sea and was possibly inundated at high tide, and a pond in Wales also with nine could have been devoid of oxygen at times. Three ponds studied by other authors and all containing ten species or more were liable to dry up. This hypothesis that large numbers of species are found in places where there is no established population is based on casual collections and more systematic work is required to confirm it. Anyone having access to a pond subject to vicissitudes of the type mentioned and to a similar but more stable one nearby might find that regular collections in both were worth making. Collections should be frequent in spring and early summer, and in autumn particularly after spells of still sunny weather, for the indications are that that is when most corixids fly. Not all, however, for Professor H. P. Moon has found corixids on the ice covering Lake District pools and the collections in the reservoir mentioned earlier showed a marked increase in the numbers of *Arctocorisa germari* between 15 December and 30 January.

Only one systematic study of recolonization is known to the author. It was made by Professor A. Thienemann between 1944 and 1947 in a small concrete basin in his garden beside a lake at

Plön in N. Germany. It was kept full of water all summer but drained during the winter. During the four years, 103 species were recorded, and of these 10 occurred in all four years, 14 in three, and 20 in two. Seventy-seven were insects to whom access posed no difficulties and the 18 species of mite probably came in attached to them. That leaves only 8 species aquatic throughout life and of these 2 were oligochaetes and 6 small crustaceans. Thirty-six species were chironomids on which difficult group Thienemann was himself an expert, and he sought assistance from experts on six of the other groups. Clearly this is not an undertaking to be recommended to anyone to whom so much expertise is not available.

However, many of the species occurred in small numbers and only 21 species were recorded as abundant, which brings work of this kind more within the compass of those not working in a freshwater research institute. The occurrence of these species is shown in Table 2.

Table 2 The number of common species in various groups found in a garden tank which was empty all winter, and the number of years in which they were recorded

| | No. of years | | | | |
	4	3	2	1	*Total*
Chironomidae	2	2	1	1	6
Chaoborus plumicornis	1	—	—	—	1
Ephemeroptera *Cloeon dipterum*	1	—	—	—	1
Coleoptera	2	—	—	—	2
Cladocera	—	—	—	3	3
Copepoda	—	—	1	—	1
Ostracoda	—	1	—	—	1
Mites	3	1	—	—	4
Corixidae	—	—	1	1	2
Total	9	4	3	5	21

The final question that Thienemann asks is about the origin of the various species. That of fourteen is unknown, twenty-eight occur in local small bodies of water, and the remaining sixty-one are known from the lake. The difference between a small concrete tank and a lake is considerable, and it seems likely therefore that the oviposition of most of the species, chironomids in particular, must be more haphazard than that of mosquitoes, which was discussed earlier.

Anyone concerned only with the common species, or even some of the common species, might take observations of this kind further. What, for example, would have been the result of keeping the tank full of water throughout the year? A concrete tank would not have been suitable as ice would have cracked it, but now glass-fibre tanks, which can expand sufficiently to avoid this disaster, are available. Poly-glass of Morecambe, Lancs, will make one of any dimensions. Would a stable community with a much smaller number of species have established itself, or would the occasional invader have upset the balance now and then? If so how large must a piece of water be before the community in it is reasonably stable? This is a question that can only be studied experimentally in containers whose water-level is maintained artificially, because small bodies of natural water are unstable physically. On 19 July 1946 Thienemann recorded small larvae of the large dragon-fly *Libellula depressa*, and on 16 September they were full grown. Their effect on the rest of the fauna was not observed. These lurking carnivores would of course not reduce a population in the same way as a hunter, as they must wait until prey comes within reach, which it will do progressively less often as its numbers decrease. Little is known about the regular movements of any freshwater animal, either predator or prey, nor about the distance at which a predator becomes aware of potential prey. If there was some information, it should be possible to calculate how long a hunting carnivore would take to extinguish a population of animals on which it fed. Rate of reproduction of the prey also comes into the equation, and another point is whether the carnivore can prevent its brothers and sisters sharing the prey. If it cannot, one oviposition by, for example, a beetle would mean certain extinction of the predator in a small volume of water.

Most animals that are preyed upon hide from their predators in vegetation or in debris. Waterfleas which inhabit the open water rely on small size and rapid reproduction to avoid excessive losses to predators. Their reproduction, however, is checked by crowding and therefore there must be a minimum volume of water in which a population could avoid extinction by a preda-

124

tor. Such animals have the advantage from the experimental point of view of being visible to the experimenter. Suitable carnivores that might be used are larvae of Coleoptera or of the phantom larva *Chaoborus*. The latter probably feeds at night but traces of a meal can often be seen in its gut through the transparent body and the number that had fed each morning in different concentrations of prey might give some preliminary information about its hunting ability.

A comparison that springs to mind at once now when so much interest is focused on conservation and the environment is between polluted and unpolluted places. The word pollution is used in two senses. The Medical Officer of Health regards as polluted any water contaminated by crude sewage and therefore liable to harbour the bacteria that cause certain diseases, but biologists generally use the word only when the contaminating substance is present in sufficient amount to affect the fauna and flora adversely. He may regard small amounts as a desirable source of enrichment. All the classic studies have been of running water. Pollution in the latter sense is due generally to organic matter, usually sewage, which is not harmful until the quantity is sufficient to lower the concentration of oxygen to an unfavourable level. A grossly polluted river in which oxygen is reduced to zero or only a little above it is carpeted by the unsightly grey wisps of sewage fungus, which is in fact a bacterium that forms filamentous sheaths. As the river recuperates downstream there are well-known stages. Tubificid worms can tolerate very low oxygen concentrations, and at higher values they are joined by chironomids. Lower down *Asellus*, snails, leeches and *Sialis* are very abundant and then the fauna gradually reverts to that of an unpolluted river. Other substances are poisonous. The classic study was carried out in Wales where it was shown that heavy metals from the spoil heaps beside old lead mines can kill all the fauna. Insects are more tolerant than animals such as flatworms and snails, and establish themselves sooner as the metals are diluted downstream.

Pollution of ponds brings us back to oxygen concentration which has been discussed several times already. It is often due to pollution by cattle but may be due to accumulation of dead

125

leaves. Pollution is therefore not entirely an 'unnatural' phenomenon, the product of modern civilization. When most of Britain was covered by forest, and rivers traversed the plains through a series of flood-channels and marsh pools, it was probably common.

A piece of standing water that is continuously and heavily polluted will contain a large population of rat-tailed maggots. Some species of mosquito are also found in such waters. *Sigara lateralis* is abundant in less grossly polluted ponds. Dr A. Pacaud in France has worked on the Cladocera of ponds. He tested the resistance of various species to oxygen lack by enclosing them in a sealed jar, waiting until about half had died, and then measuring the oxygen concentration. He found various other factors with which the success of certain species, notably *Daphnia pulex* and *Moina brachiata* in highly polluted ponds could be correlated. Their feeding limbs did not become clogged with the innumerable bacteria, as those of some other species did, and they throve on the algae without a hard cuticle which abound in such places. *Moina* was associated particularly with duckponds, and Pacaud believed that it was favoured in some way which he did not discover by the clay particles which the activities of the ducks keep in suspension.

All changes in small bodies of water are likely to be abrupt and great, and is unlikely that a series comparable with that in rivers exists. Large and therefore more stable bodies of water are preferable for studies of this kind, particularly lakes of such size that changes in the fauna can be detected with increasing distance from the source of pollution or enrichment. The Lake District lakes have been mentioned already in this context (Fig. 31).

There is even less information about the effect of other forms of pollution on the animals and plants of standing water. Ullswater contains fewer flatworms than might be expected and this could be related to the lead which, until recently, was mined in its drainage area. The concentration of this metal in the soil is high but in the water only a little above that of other lakes. Swans that establish themselves on the lake never live long and their deaths have been attributed to lead-poisoning though the author has not heard of any autopsies to establish the cause of death. Spoil from coal-tips contains sulphur among other things and this forms sulphuric acid in the water that drains from them. The effect of this has been studied in America but not in Britain. In America open-cast coal mines leave behind collections of water just as our stone quarries do, but British law demands that such excavations are filled in.

Another comparison which could be made is between ponds which dry up regularly but for different periods. Several animals are adapted to life in such places, which presumably offers the advantage of freedom from predators at least in the early stages. The fairy shrimps belong to a group of simple Crustacea which can feed only by swimming and, as they are large, they cannot counteract the effects of predation in the way their smaller relatives, the Cladocera, do.

They are therefore confined to temporary waters to which they have been able to adapt by means of an egg that will hatch only after it has been dried. Mosquitoes of the genus *Aedes* lay their eggs on damp ground and these hatch when inundated. A fact of some importance to farmers is that during the summer evaporation tends to exceed rainfall whereas in winter the reverse holds and soils are waterlogged. Temporary pools are likely, therefore, to contain water throughout the winter and to start to disappear when evaporation begins to exceed rainfall, but it is needless to state that the regime will be an irregular one in a climate such as that of the British Isles. Heavy rain lasting long enough to fill temporary pools can fall at any time during the summer. If all the mosquito eggs hatch the first time they are submerged, the resulting larvae may perish as the pool dries out before they have had time to complete development. Professor J. D. Gillett has investigated how certain African species meet this danger. He finds that eggs differ in the degree of stimulation they require before they hatch and that this is genetically controlled. It appears that selection maintains a high degree of diversity in natural populations with the result that only some eggs hatch at the first submergence. There are, therefore, some larvae to take advantage of a temporary flooding, but, if this does not last long enough for the completion of development, there remain eggs that do not hatch until later inundations. By mating adults from eggs that had hatched easily he was able to produce batches of eggs all of which hatched on slight stimulation, and by mating adults from eggs that had not hatched until after much more stimulation he could produce batches all of which had that characteristic. This work has not been repeated elsewhere and would be difficult. Few animals are easier to rear from egg to adult than mosquitoes but the adults of most species will not mate in captivity.

The main work on small bodies of water known to the author is also by a German working in the region of Plön. He records that the characteristic animals of temporary pools are fairy shrimps if they are large, and mosquito larvae if they are small. There are also some mites and a chironomid which are characteristic but whether this description applies also to various small Crustaceans that are often found is doubtful. Two snails, *Planorbis leucostoma* and *Lymnaea glabra*, are found in calcareous ponds. A French worker has recently described how the former seals its shell at the onset of drought and aestivates on the bottom of the pond.

There is much scope for systematic observations on temporary pools. The longer they last the more they are invaded by wanderers such as the beetles, whose arrival must affect the regular inhabitants adversely. There are probably other species which, though not as completely adapted to these conditions as the fairy shrimps and mosquitoes can tolerate a period of drought provided it is not too long. The bottom of a depression that holds water all winter is probably damp throughout the summer, and Hynes has shown that many stream species can survive in the damp conditions at the bottom of a waterless stream bed in the egg stage. The comparison, however, must not be stretched too far since winter growth is much commoner among stream than among pond animals.

As stress has been laid on the importance of cover, comparison of places with and without vegetation would be instructive if it were possible to find a place where the absence of vegetation was not due to some factor likely to affect the animals directly. Rooted vegetation is absent from ponds with a rock bottom, such as quarry excavations; from ponds with a sandy or gravelly bottom that is ` disturbed often enough by waves to prevent plants obtaining a roothold; from ponds so enriched with organic matter that algae prevent sufficient light reaching the bottom and from ponds in woods where trees cut off the light. It will be recollected that the vegetation in Hodson's Tarn decreased with the passage of time, a change tentatively attributed to changes in the bottom. Any comparison that is possible has been discussed already in the paragraphs devoted to changes with age. Experiments in artificial containers with artificial vegetation are a possibility.

Comparison can also be made between the communities inhabiting different kinds of vegetation. Ideally this should be done in two or more ponds each with a pure stand of one species, for in one pond any numerous species for which one kind of plant provides peculiarly suitable conditions for egg-laying or for the survival of the very young will spread to other plants as it develops.

A pond completely overgrown with *Carex, Sphagnum* and *Menyanthes* with a little *Potamogeton natans* in the more open patches provided a contrast with Hodson's Tarn. The pond is probably an old quarry and the vegetation now forms a floating mat, through which a pond-net stick two metres long can be pushed without contact with a solid bottom being made. *Pyrrhosoma* is numerous as it is in Hodson's Tarn, but *Enallagma* has not been recorded, possibly because emergent vegetation impedes the characteristic mating flight close to the surface of open water. Animals common in this pond but not in the tarn are various species of beetle and *Notonecta*. Experiments in the tarn suggest strongly that their abundance is not related directly to the vegetation but indirectly because it is so thick that fish cannot inhabit it. The presence of mosquito larvae in the pond and not in the tarn is likely to be related to the much thicker plant cover. Two species of corixid, *Hesperocorixa castanea* and *H. sahlbergi*, are numerous in the pond. The former occurs in Hodson's Tarn and many others in places where there is thick vegetation, and *H. sahlbergi* indicates that the pond is more productive than the tarn. *Sigara distincta*, and *S. scotti*, abundant in more open places in many tarns including Hodson's have not been taken in the pond.

A critical examination of the last few statements concludes

129

this chapter. The one about the corixids is based on a considerable amount of observation. The habitats of the various species can be defined with precision, though little is known about how each one is confined to the conditions in which it is found. The statement about the effects of predation is based upon an experiment and can be made with some confidence. The remainder are the sort of hypotheses that can legitimately be put forward during the course of descriptive ecology in the awareness that they are Aunt Sallies which experiment or further observation may knock down.

8 PRODUCTION

SUMMARY
Production can be calculated if the initial number of eggs, the mortality at frequent intervals, and the average weight at these intervals is known. Average weight at intervals during development is not hard to obtain, but an accurate sample from which numbers of eggs and very small young can be discovered is very difficult. When production is known, it is important to discover what it has been produced from. Good data for captive populations are available, but any calculation about a wild population must be based on an assumption of how much of the energy taken in is used in hunting and avoiding enemies.

The study of production is so much to the fore at the present time that it would be unthinkable to ignore it. The fundamental principles are simple but the difficulties in the way of obtaining accurate results are great. As Phillipson (1966) points out in his excellent little book on the subject, the manpower required for any but the simplest undertaking is not available under present conditions.

Almost every book that deals with the subject refers to K. R. Allen's work on the Horokiwi, a river in New Zealand, the reason being that this is still one of the most complete studies made. It concerns one species only, the trout, and that one particularly suitable. The Horokiwi is some ten miles long, and, by isolating stretches by means of nets at either end and then seining until no more fish were caught, Allen was able to ascertain the total population. Before they were returned to the water some of the fish were marked. The number of marked fish in a later netting bore the same relationship to the total catch as the total number of marked fish (which was of course known) bore to the total population. Marking, therefore, provided a method of checking the assessment of total population, and it also

131

revealed that the fish did not move much up or down stream, which greatly facilitated later calculations.

Samples of fish killed and examined showed the proportion of females, the size at which they first produced eggs, and the number of eggs that a fish of a given size would produce. Armed with this information, Allen could calculate the total number of eggs produced each breeding season. Mortality of eggs in the redds was found to be very low. Thereafter regular sampling revealed the losses suffered by a year class and, from total numbers and the average weight, production in terms of wet weight of fish could be calculated. Of 1000 fry emerging from a redd only fifteen were still alive at the end of six months. After one year there were seven survivors and the population continued to fall by about 50 per cent every six months till thirty months after hatching there was one left (Fig. 38b). Average weight rose steadily from 2 oz at six months to 16 oz at thirty months, by which time the original thousand had produced a total of 10 lb of fish. Allen assumed that of the fish which died in six months half were still alive midway between the two dates (Fig. 38a), an assumption which has been challenged only for the first six months. Allen assumed that mortality was as indicated by the right-hand dotted line in Fig. 38a, but it is probable that the left-hand dotted line, which shows heavier mortality in the earliest stages is nearer the truth. The contribution of small fish to the total production was, therefore, a little less than Allen calculated, but it was still substantial, the great numbers compensating for small size.

A few similar studies have been made. Crisp (1962a and b) studied *A. germari* in a Pennine reservoir, a place which offered the advantages of a uniform stony substratum along one side and a restricted fauna. He found traps unsuitable for quantitative work and had to rely on catch per standard net sweep, which presented him with the problem of translating his figures into number per unit area. It was possible to count the number of eggs per unit area on stones examined and freed of eggs once a week during the egg-laying season, which lasted from late April to mid-August.

Also known was the number that had hatched and this, assuming no mortality after hatching, was the number of

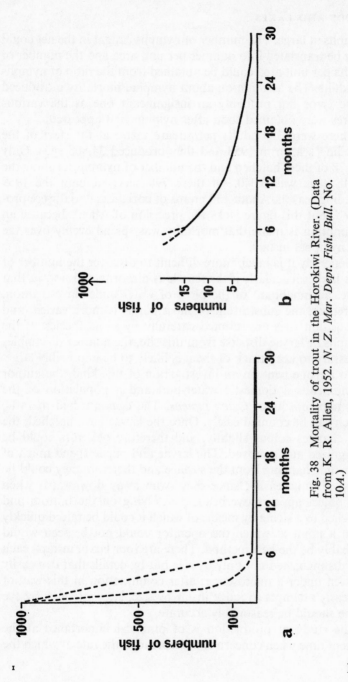

Fig. 38 Mortality of trout in the Horokiwi River. (Data from K. R. Allen, 1952. *N. Z. Mar. Dept. Fish. Bull.* No. 10*A*.)

nymphs at large. The number of nymphs caught in the net could now be translated into number per unit area and the number of adults per unit area could be obtained from the ratio of nymphs to adults. The assumption about nymphal mortality introduced some error but probably an insignificant one as the various figures were obtained soon after nymphs first appeared.

There were 380 adults per square metre at the start of the breeding season in 1958 and they produced 34,808 eggs. Only 20,178 of these hatched and the number of nymphs reaching the adult stage was 3,040. Of these 798 survived until the 1959 breeding season. About 130 grams of corixid material were produced, but this figure lacks the precision of Allen's because an assumption is made that mortality was spread evenly over the five nymphal instars.

Generally it is much more difficult to discover the number of eggs laid, particularly if the female is airborne and not, at that stage, an inhabitant of the piece of water under investigation. Moreover, the substratum of most ponds is more varied, and oviposition sites are selected carefully by some species. If the nymphs or larvae disperse from this site, the number of samples necessary to keep track of them is likely to be impossibly large.

To anyone bent on an investigation of this kind, the author recommends a domestic water-butt and a population of the common mosquito, *Culex pipiens*. The eggs are laid in rafts, which can be counted easily. Once the larvae have hatched, the raft changes colour slightly, and therefore old rafts could be recognized and removed. The larvae and pupae spend much of their time hanging from the surface and therefore they could be sampled quite easily; since they swim away downwards when any object appears overhead, a net lying on the bottom and attached to a string by means of which it could be raised quickly from a point at which the operator could not be seen would probably be the best method. There are four larval instars, each one distinguished not only by size but by details that can easily be seen under a microscope; after confirmation in this way of the early attempts to judge instar by size, later attempts by eye alone should be reasonably accurate.

The study of production is of practical importance at the present time when concern is widespread at the rate at which the human population of the world is increasing. It is important to know not only how great the production of any one species is, but what it is producing it from. There are four variables, as shown in the figure.

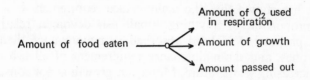

If three of the four are known, the fourth can be calculated. This has been a popular line of study in the laboratory. The work of Slobodkin on *Hydra* has been mentioned, and this small sessile carnivore is a convenient object for study by those who have not much space to work in, though a small animal cannot easily be weighed as accurately as a large one. The most popular animal has been the trout.

It will have been noticed that the imprecise word 'amount' was used in the formula. In the past the amount was generally determined in terms of weight, but this leads to some inaccuracy because, for example, a given weight of flatworm will contain more which can be used by a predator than the same weight of arthropod, whose external skeleton will probably not be digested. Recently with the development of the bomb calorimeter, amount has been expressed in terms of calories, which is the quantity of heat given out on combustion.

All four constituents of the equation can be measured easily when the subject is captive. It is fed a known amount of food and any left over is measured. The faeces are collected and their calorific value ascertained. Weight is easily measured. Oxygen consumption is the only factor in the equation that cannot be monitored continuously but occasional measurements generally show that it does not fluctuate much.

The oxygen consumed by a fish leading an uneventful indolent life in a tank is required largely for the combustion of that amount of energy which will keep the body functioning. The

135

food supplying this is referred to as the maintenance requirement. It is not much higher than the basal metabolism, which is indicated by a consumption of oxygen at which the animal can just remain alive performing no more than essential movements such as respiration. Dr M. E. Brown, a simple account of whose extensive work on trout may be found in Frost and Brown (1967), has shown that the maintenance requirement is not a fixed proportion of anything; small fish devote a relatively higher proportion of food absorbed to maintenance than do larger ones, and the maintenance requirement of all fish rises with increasing temperature. Moreover, growth is not constant, being fastest, as in most animals, in the early stages, although the trout, unlike most other animals, can continue to grow throughout life. If food is in excess, fish make less efficient use of it. Add to all this the fact that various workers have given to fish food ranging from living animals, which contain a high proportion of water, to pellets, which contain much less, and there are no grounds for surprise that figures for weight of food required to produce a given weight of fish vary greatly. Brown states that 'using dry pellets under optimal conditions in a hatchery a yield of more than one pound of trout flesh for each two pounds of food used can be maintained throughout the year'. Generally, however, it is found that the biomass produced by a later stage in the food chain is about 10 per cent of the one on which it feeds.

Anyone attempting to study wild populations is faced with various difficulties, and the only factor in the equation that can be measured easily is growth. The difficulties of finding out how much a wild animal eats have been dwelt upon earlier. Some authors, in possession of data on growth, have calculated from laboratory findings of the type just described, how much food is required to produce this, but they can only guess how much energy was used in hunting and catching this food and perhaps in fleeing from predators. In other words oxygen consumption is unknown. Dr J. H. Lawton in a series of recent papers has reported studies on *Pyrrhosoma*, an animal which, since it lurks for its prey, does not use much energy in this way. He ascertained in the laboratory the relation between amounts of different prey consumed and amounts of faeces produced, and then deduced

136

how much wild specimens had eaten from the amount of faeces they produced when taken into the laboratory. Gradually no doubt calculations based on fewer and fewer assumptions will be possible, but it is the kind of work which, involving many meticulous measurements, is likely to be taken further only by the professional. Anybody reading accounts of work of this nature is advised to examine them closely to see what assumptions are made.

Water has the advantage that animals from adjoining biotopes are not entering and leaving all the time, which makes it an easier place in which to study production at the upper end of the food chain than a terrestrial plot. Conversely at the lower end it is more difficult because so much organic and inorganic matter is entering constantly. Hodson's Tarn, for example, is fed by seepage and by one small inflow. It receives many pine needles each year. The amount of water entering the tarn varies from hour to hour on a wet day and the concentration of solutes in it varies too. Any calculation not based on continuous records would be liable to grave error. The number of dead leaves falling into the water could be ascertained by means of a receptacle of known area that retained the leaves. The amount washed in could be ascertained too by means of a technique used by Dr Crisp. He netted a Pennine stream, but found it necessary to change the nets frequently when the stream was in flood, and accordingly invented a device to set off an alarm clock whenever half an inch of rain had fallen. Decomposition and consumption of leaves could be discovered from loss of weight of the contents of a fine nylon net.

Phytoplankton in a pond such as Hodson's Tarn must be an uncertain source of food since it is liable to be diluted so greatly by heavy rain, and the epiphytic algae are probably more important at the base of the food chain. The larger the volume of water the greater the contribution of the phytoplankton to the primary production; that of the epiphytes is unimportant in a big lake. The Secchi disc, described earlier, gives a rough measure of the amount of plankton in otherwise clear water. Phytoplankton may be enclosed in a container of known volume, sedimented in some substance such as iodine, and counted, a process that requires time. Alternatively the chlorophyll may be extracted in acetone or methanol and measured in a spectrophotometer. Full details of these and other processes may be found in IBP Handbook No. 12 (Vollenweider, 1969).

Epiphytes will grow on a variety of substrata, including glass slides, and production can be measured by weighing. Allowance must be made for various complications. Artificial substrata enclosed by netting to exclude grazers must be compared with others freely accessible. Growth on clean substrata should be compared with growth on others that have been exposed for a longer period, as it is reasonable to suppose that the early colonists take steps, possibly a secretion of some kind, to prevent later comers settling on them and smothering them. This raises the question, which nobody seems to have tried to answer, of how rooted plants react to the settling of epiphytic algae. It must be disadvantageous since it cuts off the light. Algae are known to produce substances that inhibit the growth of other species and so are some animals, and it seems unlikely that higher plants do not protect themselves in this way. If they do, growth of epiphytes on artificial substrata will not reflect growth in the tarn as a whole.

The plankton is an unsatisfactory source of food in Hodson's Tarn because it is liable to be washed away by rain but even in the more stable conditions of a lake it is still unsatisfactory from the human point of view. The reasons, set out earlier, are that too many algae are too large to be eaten by any planktonic animal and all the organisms are so small that a fish must use a good deal of energy in obtaining a meal. Production would be higher if there was a bigger area of substrate on which attached algae could grow and make use of the nutrients washed into the lake. The nutrients taken up would not then be at risk of being washed down the lake outlet or carried to the bottom when the organism died. Algae would provide both food and cover for bottom-dwelling animals, which, on account of their larger size, are preferable to plankton as a food for fish. An investigation of how to increase production in this way has been started in Hodson's Tarn, and Pl. 3b (facing p. 80) shows one of the plastic squares that are suspended in midwater. Most of the species found in the tarn occur now and then on a flat square but the only regular colonists are net-spinning caddis larvae and *Lymnaea peregra*. It will be observed that the square in the illustration is pleated and this produces a marked increase in the number and variety of colonists. A mat of artificial *Littorella* (Pl. 1*b*) suspended in midwater sustains a population almost the same as that of a similar piece in a bed of real *Littorella*.

In general production is not a field recommended to anyone not well provided with facilities and with time. On the other hand it is a field in which anybody, scientist and layman alike, should be well informed. The traditional diet of the Englishman is beef, and the output of specious argument and nationalistic claptrap that would fill the columns of the newspapers and pour forth from public platforms if any suggestion were made that it should cease to be can easily be imagined, particularly at this time (November 1971) when entry into the Common Market is exciting so much controversy. The statesman of the future will, nevertheless, be giving thought to the question, pondering perhaps on Fig. 5–4 in Phillipson's book, which shows that it takes four times as long to turn hay into beef as into the same amount of rabbit flesh. The rate at which different species grow is thus one important consideration, and another is whether or not an animal has to use energy to maintain its temperature above or below that of the environment. Temperate countries must obviously depend on warm-blooded animals, which can grow during the winter. In tropical countries cold-blooded animals may be more productive. Already there is a big organization studying the best means of raising fish in tropical lands. Some botanists too are thinking well ahead. A number of years have passed since the late Professor Pearsall, in a presidential address to the British Association, pointed out that the spruce is one of the few plants which not only grows on the poor soils that cover much of our land but which grows throughout the year in contrast to many traditional crops whose growing season extends only over a few months. He suggested that scientists should be thinking about how to turn spruce into human food. If we accept that each link in a food chain turns into flesh about 10 per cent of the link before, our reliance on beef and mutton means that 1 per cent of grass is converted into human flesh. If we could convert it directly we could feed our population. Man owes his success to

139

his ability to control his environment, but until recently few gave thought to what was happening to the environment or to the consideration that man is part of it. That era is past and difficult decisions lie ahead. The chances that the right ones are taken may well depend on the number of people sufficiently well informed to be able to assess the arguments put forward by the various sides.

REFERENCES

The list contains a number of books in which the reader may find further information about any line of work which he finds of particular interest. Certain accounts of important original work, too recent to be cited in any book, are also included, together with desciptions of methods likely to be useful to anyone who contemplates work of his own.

Blunck, H. 1924. Die Entwicklung des *Dytiscus marginalis* L. vom Ei bis zur Imago. 2 Teil. Die Metamorphose (B. Das Larven und das Puppenleben). *Z. wiss. Zool.* **121**; 171–391.

Clegg, J. 1952 (3rd ed. 1965). *The Freshwater Life of the British Isles.* (Wayside and Woodland series) London: Warne. 352 pp.

Corbet, P. S., Longfield, C. and Moore, N. W. 1960. *Dragonflies.* New Naturalist, London: Collins, No. 41. xx + 260 pp.

Crisp, D. T. 1962a. Estimates of the annual production of *Corixa germari* (Fieb.) in an upland reservoir. *Arch. Hydrobiol.* **58**, 210–23.

Crisp, D. T. 1962b. Observations on the biology of *Corixa germari* (Fieb.) (Hemiptera Heteroptera) in an upland reservoir. *Arch. Hydrobiol.* **58**, 261–80.

Davies, R. W. 1969. Predation as a factor in the ecology of triclads in a small weedy pond. *J. Anim. Ecol.* **38**, 577–84.

Dendy, J. S. 1965. Use of woods to determine the depths of oxygen distribution in ponds. *Progr. Fish Cult.* **27**, 75–8.

Edwards, R. W. and Brown M. W. 1960. An aerial photographic method for studying the distribution of aquatic macrophytes in shallow waters. *J. Ecol.* **48**, 161–3.

Fox, H. M. 1921. Methods of studying the respiratory exchange in small aquatic organisms, with particular reference to the use of flagellates as an indicator for oxygen consumption. *J. Gen. Physiol.*, **3**, 563–73

Frost, W. E. and Brown, M. E. 1967. *The Trout.* New Naturalist Special Volume. London: Collins. 286 pp.

Galtsoff, P. S., Lutz, F. E., Welch, P. S., and Needham, J. G. 1959. *Culture Methods for Invertebrate Animals.* (1937 Comstock.) 1959 New York: Dover. xxxii + 590 pp.

Hackett, L. W. 1937. *Malaria in Europe. An ecological Study.* London: Oxford University Press. xvi + 336 pp.

Hynes, H. B. N. and Williams, W. D. 1965. Experiments on competition between two *Asellus* species (Isopoda Crustacea). *Hydrobiologia*, **26**, 203–10.

Kaushik, N. K. and Hynes, H. B. N. 1971. The fate of the dead leaves that fall into streams. *Arch. Hydrobiol.* **68**, 465–515.

141

Macan, T. T. 1959 (6th imp. 1970). *A Guide to British Freshwater Invertebrate Animals*. London: Longmans. x + 118 pp.

Macan, T. T. 1963. *Freshwater Ecology*. London: Longmans. x + 338 pp.

Macan, T. T. 1970. *Biological Studies of the English Lakes*. London: Longmans. xvi + 260 pp.

Macfadyen, A. 1957. *Animal Ecology Aims and Methods*. London: Pitman, xx + 264 pp.

Mackereth, F. J. H. 1963. Some methods of water analysis for limnologists. *Sci. Publ. Freshwat. Biol. Ass.* No. 21, 71 pp.

Mackereth, F. J. H. 1964. An improved galvanic cell for determination of oxygen concentration in fluids. *J. Sci. Instrum.* **41**, 38–41.

Mann, K. H. 1962. *Leeches (Hirudinea) – Their Structure, Physiology, Ecology and Embryology*. Oxford: Pergamon. x + 201 pp.

Mellanby, H. 1938 (6th ed. 1963). *Animal Life in Freshwater*. London: Methuen. xii + 308 pp.

Mortimer, C. H. and Moore, W. H. 1970 (revised ed.). The use of thermistors for the measurement of lake temperatures. *Mitt. int. Ver. Limnol.* No. 2. 42 pp.

Phillipson, J. 1966. *Ecological energetics*. Studies in Biology No. 1, Inst. Biol. London: Edward Arnold. 57 pp.

Potts, W. T. W. and Parry, G. 1964. *Osmotic and Ionic Regulation in Animals*. London: Pergamon. xiv + 423 pp.

Reynoldson, T. B. 1966. The distribution and abundance of lake-dwelling triclads – toward a hypothesis. *Adv. ecol. Res.* **3**, 71 pp.

Slobodkin, L. B. 1962. Predation and efficiency in laboratory populations. In *The exploitation of natural animal populations*. Oxford: Blackwell. 223–41 pp. *Symp. Br. Ecol. Soc.* 2.

Thienemann, A. 1954. Ein drittes biozönotisches Grundprinzip. *Arch. Hydrobiol.* **49**, 421–2.

Vollenweider, R. A. (ed). 1969. *A Manual on Methods for Measuring Primary Production in Aquatic Environments*. IBP Handbook No. 12. Philadelphia: Davis. xvi + 213 pp.

Wynne-Edwards, V. C. 1962. *Animal Dispersion in Relation to Social Behaviour*. Edinburgh: Oliver & Boyd. xi + 653 pp.

INDEX

acetone, 138
adaptation, 46–53
Aedes, 127
Aeshna, 28
Agrion, 84, 109
  splendens, 91, 92
  virgo, 91, 92
algae, 16, 17, 20, 23, 25, 61, 73, 116, 117, 129, 138
  decomposition of, 58
  epiphytic, 137, 138
  food, as, 26, 126
Allen, K. R. 131, 132, 133, 134
aluminium, 61
Anax, 39
  imperator, 33
Ancylus, 94
  fluviatilis, 93
  lacustris, 93
Anodonta (swan-mussel), 27
Anopheles, 89, 92
  maculipennis, 68, 69, 70, 87
  maculipennis/typicus, 69
  minimus, 84, 90, 91
  sacharovi, 69
Aphelocheirus, 53
Arctocorisa carinata, 83
  (*Corixa*) *germari*, 37, 38, 122, 132
Artemia salina, 117
Asellus, 73, 105, 106, 121, 125
  aquaticus, 97
  competition in, 97
  meridianus, 97
  as prey, 96
  in Windermere, 81–3
Asterionella, 17
Azolla, 25

backswimmer (*Notonecta*), 29
bacteria, 18, 26, 73, 103, 105, 116, 125, 126
Baetis rhodani, 39
Baikal, Lake, 114

bathymetrical survey of the Scottish lochs, 21
beaked sedge (*Carex rostrata*), 76
beetles, 27, 28, 47, 50, 52, 53, 80, 98, 106, 128, 129
  flour, 97
behaviour, 98, 111
Bentham, 12
bicarbonate, 16, 17, 59
biocoenotics, laws of, 50, 84
Birge-Ekman grab, 62, 63
bladder snail (*Physa*), 105
Blunck, H., 74
Bodo sulcatus, 48
bomb calorimeter, 135
borrow-pits, 90
Boycott, A. E., 119, 120, 121
Branchiura sowerbyi, 105
brine shrimp, 117
Brinkhurst, R., 67
Brown, M. E., 111, 136
  M. W., 55
Bufo, 98
bugs, 47, 52, 53, 73, 106, 121
bulrush, 26
Buttermere, 20

caddis, 27, 28, 65, 80, 81, 82, 139
calcium, 16, 17, 54, 59, 60, 94, 95, 96, 106, 107
Callicorixa wollastoni, 83
Carboniferous period, 15
Carex, 79, 87, 99, 129
  artificial, 66
  rostrata, 76
Carp, 28
cercariae, 102
chance, 118
Chaoborus, 125
  plumicornis, 123
chemical factors, 16, 17, 55, 58–61, 90
cherry, 59
chestnut, 59

143

phytoplankton, 16, 57, 62, 137
Pisidium (pea-mussel), 27, 68
planarian, 98
plankton, 62, 111, 138
  net, 20, 62
Planorbis albus, 60
  corneus, 103, 120
  leucostoma, 128
plant grab, 61
plastron, 52, 53
Plecoptera, 44, 46, 80, 81
poisons, 125
pollution, 16, 52, 109, 115, 125, 126
Polycelis nigra, 94, 95, 96
  tenuis, 94, 95, 96
Polycentropodidae, 27
Poly-glass, 124
pondweed, floating, 76
Potamogeton alpinus, 76
  natans, 76, 87, 88, 100, 129
potassium, 16, 17, 59
  permanganate, 23
Potts, W. T. W., 46
powan, 29
Pratt, D. M., 116
predation, 98, 99, 102–4, 111, 112,
  117, 124, 125, 127
production, 15–18, 103, 106, 134
  primary, 16, 137
profile of lakes, 18, 19, 20
Protozoa, 48, 73
Ptychopteridae, 52
Pyrrhosoma nymphula, 34, 35, 36, 72,
  78, 87, 99, 102, 109, 113, 129, 136

quillwort (*Isoetes*), 26

rainfall, 55
ramshorn, large, 120
Rana, 98
Rantus, 98
rat-tailed maggot, 51, 126
recolonization, 122
redd, 111, 132
rediae, 102
reed, 26
reedmace, 26
respiration, 48–53, 135
Reynoldson, T. B., 73, 94, 95, 96
Rhithrogena semicolorata, 30, 31,
  39
Ross, R., 89

rotifers, 58
Russell-Hunter, W., 37, 39, 40, 41, 49

Salmo trutta, 109
salmon, 77
  Pacific, 108, 110, 111
sampling, 54
Saw cylinder sampler, 63, 64
Scottish lochs, bathymetrical survey
  of, 21
Secchi disc, 57, 137
sedge, 26
seine net, 67, 131
sewage, 11, 107, 125
sewage disposal, 16
  fungus, 125
  Royal Commission on, 11
  works, 81
shrimp, brine, 117
  freshwater, 45, 82
  fairy, 127, 128
Sialis, 79, 80, 125
silica, 17
siliceous skeleton, 17
Sigara distincta, 83, 129
  lateralis, 119, 120, 126
  limitata, 83, 120
  nigrolineata, 83, 119, 120
  scotti, 83, 87, 129
  selecta, 83
  stagnalis, 83
Slack, H. D., 29
Slobodkin, L. B., 117, 135
snails, 89, 105, 108, 125, 128
  and calcium, 59, 60, 106, 107
  collecting, 65
  control of numbers in, 111
  culture of, 115, 116
  feeding of, 26, 103
  as food, 79, 96, 102
  life histories of, 39–41
  numbers of, 102
  origin of freshwater, 46
sodium, 16, 59
Sparganium, 93
spawning period, 37
spectrophotometer, 138
sphagnum, 74, 99, 129
springtails, 67
stability, 24
Stenelmis canaliculata, 53
Sterile mud, 76

147

Habitat Series
Other titles in preparation

Saltmarshes
Heaths and Moors